Limit State Design of Reinforced Concrete Structures

A HALSTED PRESS BOOK

JOHN WILEY & SONS
New York—Toronto

First published by
Surrey University Press
Kingswood House, Heath and Reach, Leighton Buzzard, Beds LU7 0AZ
and 450 Edgware Rd., London W2 1EG

First published 1974

Published in the U.S.A. and Canada by Halsted Press, a Division of John Wiley & Sons, Inc., New York

Library of Congress Cataloging in Publication Data

Beckett, Derrick.
Limit state design of reinforced concrete structures.

"A Halsted Press book."
1. Reinforced concrete construction. 2. Plastic analysis (Theory of structures) I. Title.
TA683.2.B4 1975 624'.1834 74-20975
ISBN 0-470-06128-6

Printed in Great Britain

Acknowledgements

In the preparation of *Reinforced Concrete* the author has again leant heavily on the originality and industry of others. Of particular inspiration have been the writings and/or constructions of R. Maillart, E. Freyssinet, G. P. Manning, A. L. L. Baker, R. H. Wood and also the numerous technical papers produced by the Cement and Concrete Association. Tables and data, from CP 110: Part 1: 1972: *The Structural Use of Concrete* Part 1, *Design Materials and Workmanship*, are reproduced by permission of the British Standards Institution, 2 Park Street, London, W1A 2BS.

From CP 110: p. 69, Fig 3.33
p. 71, Fig 3.36
p. 74, Table 3.1
p. 75, Table 3.2
p. 78, Table 3.3
p. 79, Table 3.4 and 3.5
p. 86, Table 4.1
p. 90, Table 4.2 and 4.3

Contents

Symbols vii

Preface ix

1 A Survey of Elastic Theory 1
Introduction 1
Section analysis 2
Member analysis 15

2 The Development of Limit State Analysis 23
Introduction 23
Characteristic strength 32
Design strengths 33
Characteristic load 33
Design load 33
Ultimate limit state 34
Serviceability limit states 36

3 Ultimate Limit States 40
Member analysis 40
Section analysis 68
Summary 82

4 Serviceability Limit States 84
Introduction 84
Calculation of widths of cracks 85
Deflection 89
Calculations of deflections 91

5 Typical Calculations 96

Appendix Reinforcement Areas 127
References 130
Index 133

Symbols

A_c	area of concrete
A_s^1	area of compression reinforcement
A_s	area of tension reinforcement
A_{sv}	cross sectional area of two legs of a link
a	deflection
b	width of section
d	effective depth of tension reinforcement
d^1	depth of compression reinforcement
E_c	static secant modulus of elasticity of concrete
E_s	modulus of elasticity of steel
F_k	characteristic load
f_{cu}	characteristic concrete cube strength
f_k	characteristic strength
f_y	characteristic strength of reinforcement
G_k	characteristic dead load
h	overall depth of section in plane of bending
M	bending moments due to ultimate loads
M_u	ultimate moment of resistance
Q_k	characteristic imposed load
V	shear force due to ultimate load
V_c	ultimate shear resistance of concrete
v	shear stress
v_c	ultimate shear stress in concrete
v_t	torsional shear stress
v_{tu}	ultimate torsional shear stress
x	neutral axis depth
z	lever arm
αe	modular ratio

Preface

This book aims at introducing students and practising engineers to the application of limit state theory to the design of the more common forms of reinforced concrete construction, the slab, beam and column. From discussions with a number of engineers it has become apparent to the author that there is a reluctance to apply limit state principles as developed in CP 110 : 1972 to the design of concrete structures, the belief being that substantially more design time is involved with little or no saving in materials. On the occasions that the author has applied limit state theory to the design of concrete structures this was certainly the case, but the primary cause for the increase in design time was due to the effort involved in satisfying the checking authorities that the method was adequate as it was not covered by the building regulations.

However, the primary concerns of the engineer are strength and serviceability and thus the introduction of CP 110 which focuses attention very sharply on both must surely be welcomed – at least in spirit. It is the author's view that the introduction of CP 110 allows the designer to channel his thoughts more directly onto considerations of strength and serviceability, to utilize more readily increasing knowledge of the properties of materials, the correlation of statistical data with regard to loading and the latest analytical techniques. Initially the design time will be greater due to lack of familiarity but eventually with the aid of computers, charts and simplified procedures the situation should improve, but the danger will always remain that juggling with arithmetic will cloud engineering judgement. From a teaching standpoint the concept of multiple limit states forms a logical approach to design but in civil engineering there is little point in developing a theory if it is not going to be applied.

1

A Survey of Elastic Theory

'Our knowledge of reinforced concrete structures has been very seriously hindered and obscured by the development of the theory of the elastic structure. The absolute mathematical certainty of the figured results when the material is assumed to be elastic (in the full scientific meaning of the term) appears to have fascinated many academical writers on the subject and has completely blinded them to the fact that the resemblance between a concrete structure and a perfect elastic structure is very slight. The only logical theory to apply to a reinforced concrete design is that which is based on the conditions which exist in an actual structure when it has just reached its ultimate load, and there are very strong indications that such a theory will be evolved during the next twenty or thirty years.'

G. P. Manning[1] (March 1924)

INTRODUCTION

Although 50 years have elapsed since the first edition of G. P. Manning's excellent book[1] was published, the latest edition[2] of the British Standard Code of Practice for the Structural Use of Concrete CP 110 : Part 1 : 1972 still, in general, recommends an elastic analysis to determine the distribution of force actions in a structure at ultimate conditions. It is only with regard to the design of solid slabs that specific reference is made to assessing the distribution of force actions by means of an inelastic approach – for example Johansen's yield line method or Hillerborg's strip method. Further, at service load conditions an elastic analysis is recommended for both member and section analysis. Thus it can be seen that classical elastic theory still forms an important part of the design process for reinforced concrete structures, and the object of this introductory chapter is to survey briefly the elastic approach to the section and member analysis.

SECTION ANALYSIS

1 FLEXURE

Hooke's law and Bernoulli's theorem from the basis of classical elastic theory for the composite material – concrete and reinforcement – and the assumptions made in the analysis of reinforced concrete sections subjected to flexure are as below:

(a) The steel and concrete stresses are within the elastic range from zero to service load.
(b) The resistance of the concrete to tensile stress is negligible.
(c) There is no relative movement of the reinforcement in the concrete (bond slip).
(d) Plane cross sections of members before loading remain plane under load.

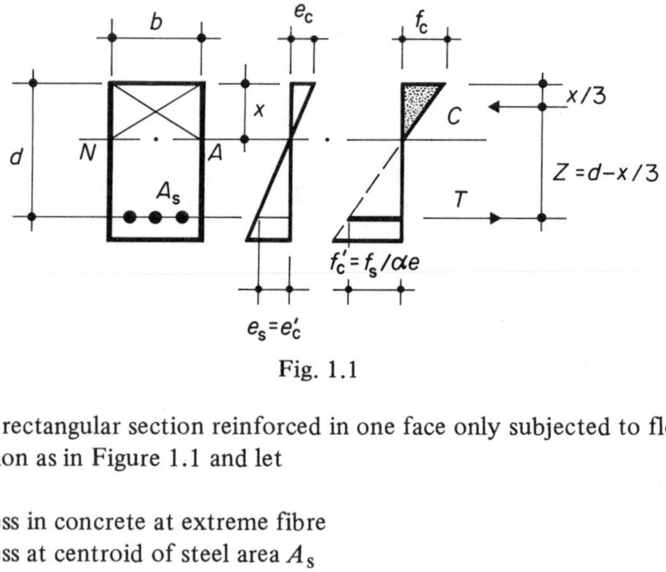

Fig. 1.1

Consider a rectangular section reinforced in one face only subjected to flexure with notation as in Figure 1.1 and let

f_c = stress in concrete at extreme fibre
f_s = stress at centroid of steel area A_s
$p = \frac{A_s}{bd}$ (steel percentage $= \frac{A_s}{bd} \cdot 100$)

At the level of the steel the strain in the steel e_s equals the strain in the concrete e_c'

$$e_s = e_c'$$

$$\frac{f_s}{E_s} = \frac{f_c'}{E_c}$$

thus $$f_c' = \frac{f_s}{\alpha e}$$

where $\alpha e = \frac{E_s}{E_c} = \frac{\text{modulus of elasticity of steel}}{\text{modulus of elasticity of concrete}}$

Referring to the stress diagram and using similar triangles (Figure 1.1),

$$\frac{f_c}{x} = \frac{f_s/\alpha e}{d-x}$$

$$\text{thus} \quad \frac{f_c}{f_s/\alpha e} = \frac{x}{d-x}$$

and putting $x = nd$

$$= \frac{n}{1-n} \quad \ldots\ldots 1(\text{i})$$

For longitudinal equilibrium the total tension equals the total compression

$$C = T$$

$$\frac{f_c}{2} bx = A_s f_s$$

$$= p \frac{bd}{100} f_s$$

$$\frac{f_c}{f_s} \frac{x}{d} = \frac{p}{50}$$

From equation 1(i)

$$\frac{f_c}{f_s} = \frac{1}{\alpha e} \frac{x}{d-x}$$

$$\text{thus} \quad \frac{1}{\alpha e} \frac{x}{d-x} \frac{x}{d} = \frac{p}{50}$$

$$p = \frac{50}{\alpha e} \frac{x^2}{d-x} \frac{1}{d}$$

Again putting $x = nd$ the above equation simplifies to

$$p = \frac{50}{\alpha e} \frac{n^2}{1-n} \quad \ldots\ldots 1(\text{ii})$$

From equation 1(ii) it can be seen that the depth to the neutral axis depends on the percentage of reinforcement and the modular ratio. The relationship between p, αe and $n(n = x/d)$ based on equation 1(ii) is shown in Table 1.1. The resistance moment of the section may be expressed in the form

$$Cz = Tz$$

It is normal to proportion a section so that the steel and concrete reach their maximum permissible stresses simultaneously; thus the ratio f_c/f_s is known and for a given value of αe equation 1(i) may be used to determine the value of n. This is referred to as a balanced section.

The resistance moment RM $= Cz$

$$= \frac{f_c}{2} bnd \left(d - \frac{nd}{3}\right)$$

$$= \frac{f_c}{2} n \left(1 - \frac{n}{3}\right) bd^2$$

$$= Q\,bd^2$$

where $Q = \frac{f_c}{2} n \left(1 - \frac{n}{3}\right)$ ..... 1(iii)

and is known as the resistance moment factor.

Table 1.1 The relationship between p, αe and n based on Equation 1(i)

	Percentage of steel								
$\alpha e + n$	*5*	*10*	*15*	*20*	*25*	*30*	*35*	*40*	*45*
0.00	0.000	0.000	0.000	0.000	0.000	0.000	0.000	0.000	0.000
0.02	0.004	0.002	0.001	0.001	0.001	0.001	0.001	0.001	0.000
0.04	0.017	0.008	0.006	0.004	0.003	0.003	0.002	0.002	0.002
0.06	0.038	0.019	0.013	0.010	0.008	0.006	0.005	0.005	0.004
0.08	0.070	0.035	0.023	0.017	0.014	0.012	0.010	0.009	0.008
0.10	0.111	0.056	0.037	0.028	0.022	0.019	0.016	0.014	0.012
0.12	0.164	0.082	0.055	0.041	0.033	0.027	0.023	0.020	0.018
0.14	0.228	0.114	0.076	0.057	0.046	0.038	0.033	0.028	0.025
0.16	0.305	0.152	0.102	0.076	0.061	0.051	0.044	0.038	0.034
0.18	0.395	0.198	0.132	0.099	0.079	0.066	0.056	0.049	0.044
0.20	0.500	0.250	0.167	0.125	0.100	0.083	0.071	0.063	0.056
0.22	0.621	0.310	0.207	0.155	0.124	0.103	0.089	0.078	0.069
0.24	0.758	0.379	0.253	0.189	0.152	0.126	0.108	0.095	0.084
0.26	0.914	0.457	0.305	0.228	0.183	0.152	0.131	0.114	0.102
0.28	1.089	0.544	0.363	0.272	0.218	0.181	0.156	0.136	0.121
0.30	1.286	0.643	0.429	0.321	0.257	0.214	0.184	0.161	0.143
0.32	1.506	0.753	0.502	0.376	0.301	0.251	0.215	0.188	0.167
0.34	1.752	0.876	0.584	0.438	0.350	0.292	0.250	0.219	0.195
0.36	2.025	1.013	0.675	0.506	0.405	0.338	0.289	0.253	0.225
0.38	2.329	1.165	0.776	0.582	0.466	0.388	0.333	0.291	0.259
0.40	2.667	1.333	0.889	0.667	0.533	0.444	0.381	0.333	0.296
0.42	3.041	1.521	1.041	0.760	0.608	0.507	0.434	0.380	0.338
0.44	3.457	1.729	1.152	0.864	0.691	0.576	0.494	0.432	0.384
0.46	3.919	1.959	1.306	0.980	0.784	0.653	0.560	0.490	0.435
0.48	4.431	2.215	1.477	1.108	0.886	0.738	0.633	0.554	0.492
0.50	5.000	2.500	1.667	1.250	1.000	0.833	0.714	0.625	0.556
0.52	5.633	2.817	1.878	1.408	1.127	0.939	0.805	0.704	0.626
0.54	6.339	3.170	2.113	1.585	1.268	1.057	0.906	0.792	0.704
0.56	7.127	3.564	2.376	1.782	1.425	1.188	1.018	0.891	0.792
0.58	8.010	4.005	2.670	2.002	1.602	1.335	1.144	1.001	0.890
0.60	9.000	4.500	3.000	2.250	1.800	1.500	1.286	1.125	1.000
0.62	10.116	5.058	3.372	2.529	2.023	1.686	1.445	1.264	1.124

Equation 1(iii) may also be derived from the equation of bending

$$\frac{M}{I_c} = \frac{f}{y}$$

where M = bending moment
I_c = second moment of area about neutral axis
f = stress at level considered
y = distance from neutral axis to level considered.

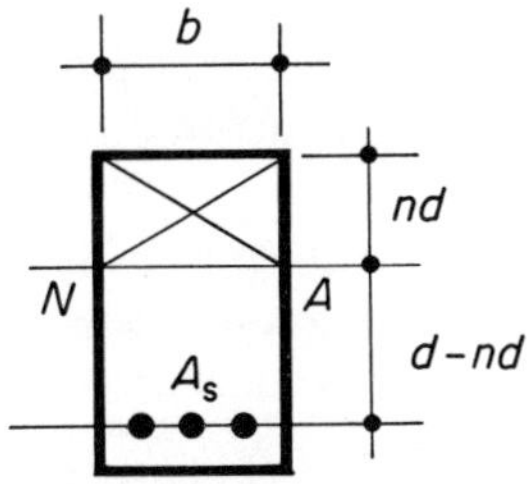

Fig. 1.2

Working in terms of the concrete, the reinforcement is replaced by an area $\alpha e A_s$ of concrete. Taking moments about the neutral axis (Figure 1.2),

$$mA_s d(1 - n) = \frac{1}{2} bn^2 d^2 \qquad \ldots\ldots 1(\text{iv})$$

The second moment of area I_c about the neutral axis NA is

$$I_c = mA_s d^2 (1 - n)^2 + \frac{1}{3} bn^3 d^3 \qquad \ldots\ldots 1(\text{v})$$

from equations 1(iv) and 1(v)

$$I_c = \frac{1}{2} bn^2 d^2 d(1 - n) + \frac{1}{3} bn^3 d^3$$

$$= \frac{1}{2} bn^2 d^3 - \frac{1}{6} bn^3 d^3$$

$$= \frac{1}{2} bn^2 d^3 \left(1 - \frac{n}{3}\right)$$

now $\quad \dfrac{f_c}{nd} = \dfrac{M}{I_c}$

thus $\quad M = \dfrac{1}{2} f_c n \left(1 - \dfrac{n}{3}\right) bd^2$

This is identical to equation 1(iii).

Where the structural depth of a member is limited, the applied moment may be greater than the resistance moment of a singly reinforced section. The resis-

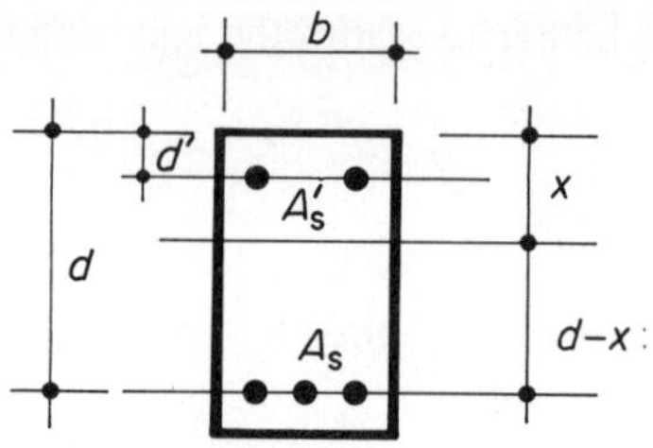

Fig. 1.3

tance moment of the section may be increased by introducing steel in both faces whilst maintaining the same effective depth d. From Figure 1.3

$$\text{RM} = Qbd^2 + A_s' f_c \frac{x - d'}{x} (\alpha e - 1)(d - d') \quad \ldots\ldots 1(\text{vi})$$

For a given moment, equation 1(vi) can be used to determine A_s'

$$\text{the area of tension steel } A_s = \frac{\text{applied bending moment}}{f_s \times \text{lever arm}} \quad \ldots\ldots 1(\text{vii})$$

As an approximation the lever arm z for the singly reinforced section may be used.

2 FLEXURAL RIGIDITY

An important property of a reinforced concrete section relating to its deformation under load is its flexural rigidity (EI). At this stage an expression for the flexural rigidity of the cracked section will be derived from the equation of bending. The significance of this equation in estimating the deflection of reinforced concrete members will be discussed in Chapter 4.

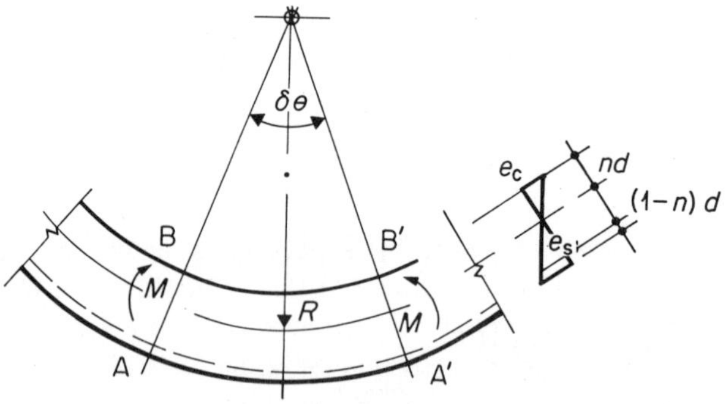

Fig. 1.4

From Figure 1.4 the strain at the level of the steel is given by

$$e_s = \frac{(R + (1 - n)d)d\theta - Rd\theta}{Rd\theta}$$

$$= \frac{(1-n)d}{R}$$

From the bending equation $\frac{M}{I} = \frac{E}{R}$

$$MR = EI$$

$$\text{and } \frac{1}{R} = \frac{e_s}{(1-n)d}$$

$$\text{thus } EI = \frac{M(1-n)d}{e_s}$$

$$\text{now } M = A_s f_s \left(1 - \frac{n}{3}\right) d$$

$$\text{and } p = \frac{A_s}{bd}$$

$$\text{hence } M = pbf_s \left(1 - \frac{n}{3}\right) d^2$$

$$\text{and } EI = pbf_s \left(1 - \frac{n}{3}\right) d^2 \frac{(1-n)d}{e_s}$$

$$\text{elastic modulus for steel } E_s = \frac{f_s}{e_s}$$

$$\text{hence } EI = pE_s \left(1 - \frac{n}{3}\right) (1-n)bd^3$$

$$\text{or } EI = Kbd^3 \qquad \ldots\ldots 1(\text{viii})$$

$$\text{where } K = pE_s \left(1 - \frac{n}{3}\right) (1-n)$$

This expression ignores the stiffening effect of the concrete in the tension zone which will increase the value of EI (see Chapter 4). Equation 1(viii) can be used to demonstrate the effect of creep on the value of the flexural rigidity of a section. The effect of sustained loading is to reduce the effective modulus of the concrete due to creep and thus increase the value of αe. It can be shown that by substituting suitable values in equations 1(ii) and 1(viii) the flexural rigidity is only moderately affected by a large reduction in the value of the modulus of elasticity for concrete. Consider the following example:

(a) $p = 1.0$ (percentage)
    $\alpha e = 10$

From equation 1(ii), $\quad 1.0 = \frac{50}{10} \frac{n^2}{1-n}$ (or Table 1.1) thus $$n = 0.36$$

From equation 1(viii),

$$EI = 0.01\left(1 - \frac{0.36}{3}\right)(1 - 0.36)E_s bd^3$$

$$= 0.00564\, E_s bd^3$$

(b) to allow for creep take $\alpha e = 30$

From equation 1(ii), $$1.0 = \frac{50}{30}\frac{n^2}{1-n} \quad \text{(or Table 1.1)}$$

$$n = 0.53$$

From equation 1(viii),

$$EI = 0.01\left(1 - \frac{0.53}{3}\right)(1 - 0.53)E_s bd^3$$

$$= 0.00387\, E_s bd^3$$

$$\text{the ratio of } EI \text{ values} = \frac{0.00387}{0.00564} = 0.687$$

Although the effective modulus of elasticity of the concrete has reduced to one third of its initial value, that is αe has increased from 10 to 30, the value of EI has only reduced to 0.687 of its initial value.

3 DIAGONAL TENSION

Consider a plain concrete beam of rectangular cross section subjected to uniformly distributed loading as in Figure 1.5(a). This loading will produce shearing and bending type force actions (Figure 1.5(b)) resulting in stress distributions at midspan, quarter span and near the supports, as shown in Figure 1.5(c). From these stress distributions the crack pattern for the beam can be predicted, the cracks being at right angles to the direction of principal tensile stress due to the interaction of f and v Figure 1.5(d). The conventional treatment of shear is to consider a section such as 1–1 in Figure 1.5(e), then at the neutral axis $f = 0$ and from the equilibrium requirements for element ABCD (a unit cube)

$$v_1 1.1 = v_2 1.1$$

$$\text{that is} \quad v_1 = v_2 = v \quad \text{(complementary shear stresses)}$$

$$\text{again} \quad v_t\sqrt{2} = \frac{v}{\sqrt{2}} + \frac{v}{\sqrt{2}}$$

$$\sqrt{2}v_t = \sqrt{2}v$$

$$v_t = v$$

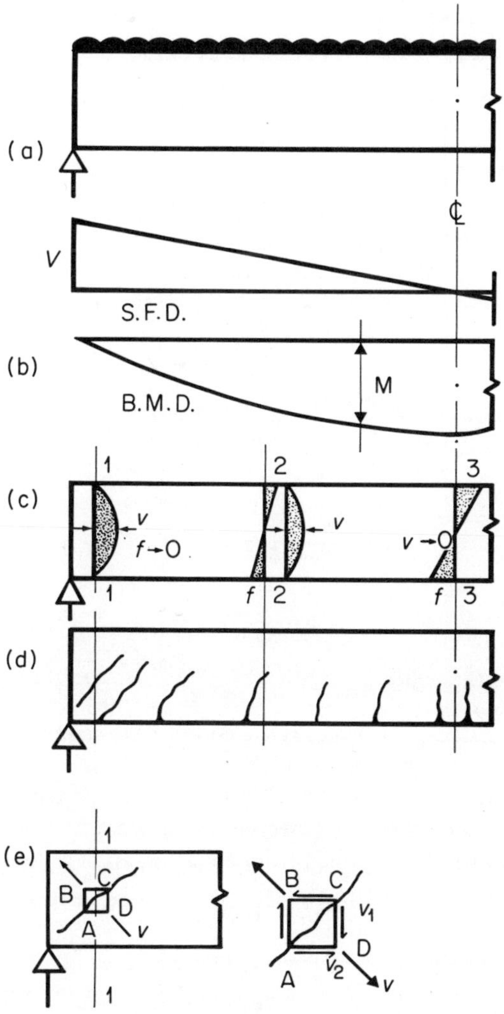

Fig. 1.5

Thus at the neutral axis the principal tensile stress is equal to v and the potential diagonal crack is at 45° to the longitudinal axis of the beam – diagonal tension. The above treatment would be reasonable for a homogeneous elastic material with equal resistance to tension and compression. For a concrete section the distribution of shear stress above the neutral axis may be assumed to take the normal parabolic form (Figure 1.6). Below the neutral axis it is assumed that the change in tension $\mathrm{d}T$ over an elemental length $\mathrm{d}x$ is resisted by a shear force on A–A.

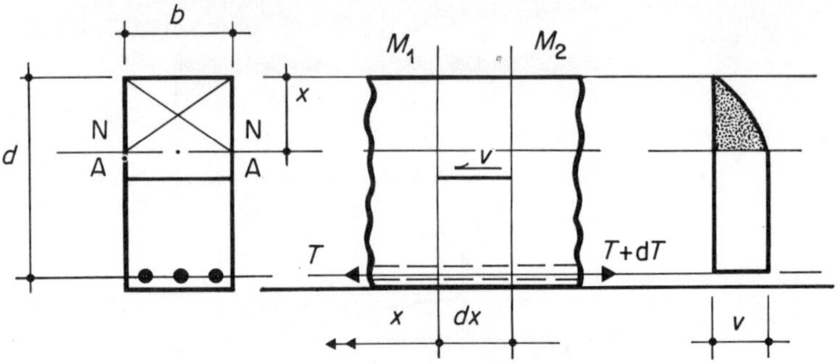

Fig. 1.6

$$\text{thus} \quad dT = vb\,dx$$

$$\text{also} \quad \frac{dM}{z} = \frac{M_2 - M_1}{z} = dT$$

$$\text{hence} \quad \frac{dM}{dx} = vbz$$

$$v = \frac{V}{bz} \qquad \ldots\ldots 1\text{(ix)}$$

where $\quad V$ = shear force at section considered

The derivation of equation 1(ix) ignores the equilibrium requirement of complementary shear stresses. The procedure in CP 114 : 1965[3] is to allow shear stresses up to about 0.9 N/mm^2 at service load. If this value is exceeded, reinforcement is provided to resist the total shear. Irrespective of the amount of shear reinforcement provided the value of v obtained from equation 1(ix) is limited to four

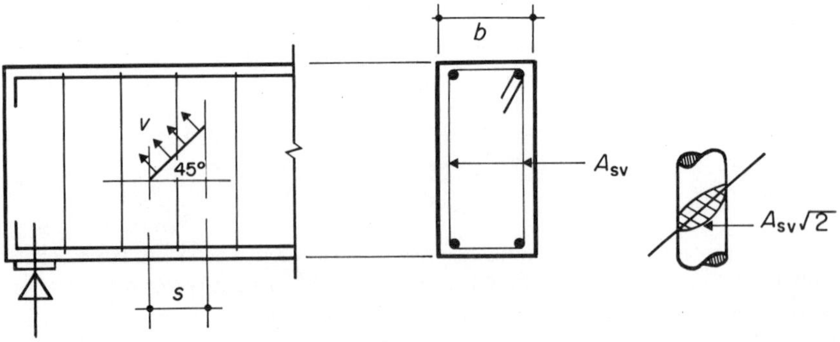

Fig. 1.7

times the permissible value. The area of shear reinforcement required is calculated using the truss analogy (Figure 1.7). Assuming a 45° crack the length between the vertical links passing through the crack is $s\sqrt{2}$.

$$\text{thus shear} = s\sqrt{2}bv$$

$$= s\sqrt{2}b\frac{V}{bz}$$

If this shear is to be resisted by the reinforcement then

$$Asv\sqrt{2}f_s = \frac{s\sqrt{2}V}{z}$$

$$s = \frac{Asvf_sz}{V} \qquad \ldots\ldots 1(x)$$

The above approach considers shear independently of bending and although safe for the case where full shear coverage is provided by closely spaced links ($s > 0.5d$), is not safe for conditions in which the shear stress v is just below the permissible value and nominal links only are provided. The resistance of a reinforced concrete section to shear is influenced by a number of factors – strength of the concrete, presence of bending moment, percentage of longitudinal steel, the beam span and type of loading; a more rational treatment of the problem is given in CP 110 (see Chapter 3).

4 BOND

The transfer of tension from the concrete to the steel is dependent on adequate bond between the steel and the concrete. If the reinforcing bars slip the full flexural and shear strength of the concrete will not develop.

(a) *Flexural bond*

It is necessary to check the local bond stress due to change in tension in the reinforcement over a short length dx. The maximum rate of change of tension $(T_2 - T_1)/\mathrm{d}x$ will occur when the slope of the bending moment diagram is a maximum. For a simply supported beam the rate of change of tension will be a maximum at the supports (for uniformly distributed loading), as in Figure 1.8. Consider two sections a distance dx apart – the difference in tension is $T_2 - T_1$.

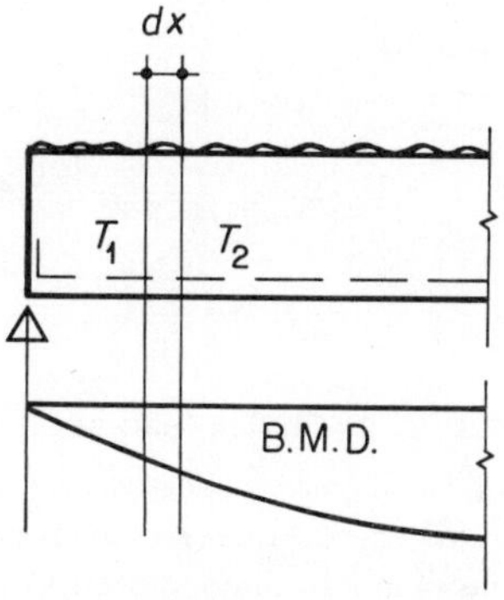

Fig. 1.8

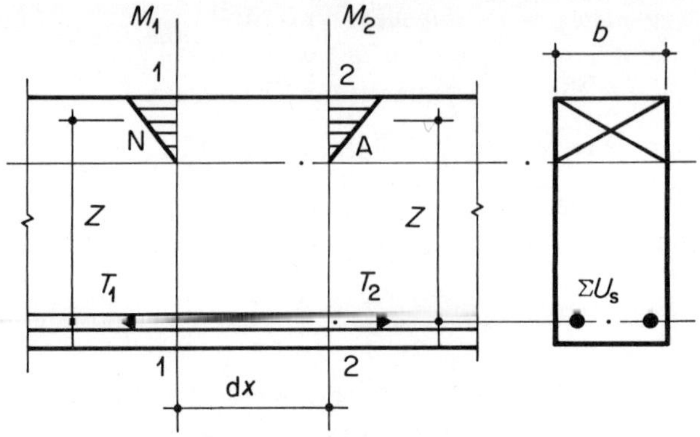

Fig. 1.9

Let f_{bs} = local bond stress between steel and concrete

ΣU_s = the sum of the effective perimeters of the tension reinforcement

Then for equilibrium (Figure 1.9),

$$\Sigma U_s f_{bs} \mathrm{d}x = T_2 - T_1$$

$$= \frac{M_2 - M_1}{z}$$

$$\text{thus} \quad f_{bs} = \frac{M_2 - M_1}{\mathrm{d}x} \frac{1}{z \Sigma U_s}$$

$$= \frac{V}{\Sigma U_s z} \qquad \ldots\ldots 1(\mathrm{xi})$$

where V is the shear force at the section considered under service load.

(b) *Anchorage bond*

To prevent bond failure the calculated tension or compression in any bar at any section must be developed on each side of the section by an appropriate embedment length or other end anchorage. For a bar of diameter d stressed to f_s the total force in the bar is $(\pi/4)d^2 f_s$.

If f_{ba} is the average bond stress over a length L, then

$$\frac{\pi}{4} \mathrm{d}^2 f_s = f_{ba} \pi d L$$

$$L = \frac{d}{4} \frac{f_s}{f_{ba}} \qquad \ldots\ldots 1(\mathrm{xii})$$

Permissible values of f_{bs} and f_{ba} at service load are given in CP 114.[3]

5 TORSION

Concrete will fail in torsion when the tensile component of the torsional shear reaches its maximum value. Saint Venant's memoir on torsion presented in 1853 to the French Academy of Science still forms the basis for torsion studies, but more recently equations have been developed using the plastic concept of concrete behaviour, and these will be derived in Chapter 3.

Returning to the elastic theory it should be noted that plane sections become warped under a torsional force action. This can be demonstrated by dividing the surfaces of a rectangular block of rubber into squares and twisting it. The maximum distortion will occur in the squares at the mid depth of the longer sides of the block. Saint Venant obtained a solution for the maximum torsional shear stress at the middle of the longer side given by

$$v_t = \frac{T}{\alpha_1 x^2 y} \qquad \ldots\ldots 1(\text{xiii})$$

where T is the service load torque at the section considered and α_1 is a constant relating to y/x ($y > x$).

The maximum stress at the middle of the shorter side is given by

$$v_t = \frac{T}{\alpha_2 x y^2} \qquad \ldots\ldots 1(\text{xiv})$$

The constants α_1 and α_2 are given in Table 1.2. It should be noted that transverse and torsional shear stresses are additive and thus limitations must be imposed on the sum of these stresses $v + v_t$.

The general form of the equations for torsional shear reinforcement may be obtained by reference to Figure 1.10(a) and (b). The applied torque T can be equated to the resistance provided by the links which may be assumed to be smeared along the length of the section to form a thin walled tube of constant thickness t. From Figure 1.10(b) the resistance of the tube provided by the constant shear flow q is equated to the applied torque.

$$\text{thus} \quad T = q y_1 \frac{x_1}{2} 2 + q x_1 \frac{y_1}{2} 2$$
$$= 2 q x_1 y_1$$

Table 1.2 Value of torsion constants α_1 and α_2

y/x	α_1	α_2
1.0	0.208	0.208
1.2	0.219	0.196
1.4	0.227	0.184
1.6	0.234	0.174
1.8	0.240	0.164
2.0	0.246	0.155
2.5	0.258	0.135
3.0	0.267	0.12
5.0	0.292	0.079
10.0	0.312	0.042
100.0	0.331	
∞	0.333	

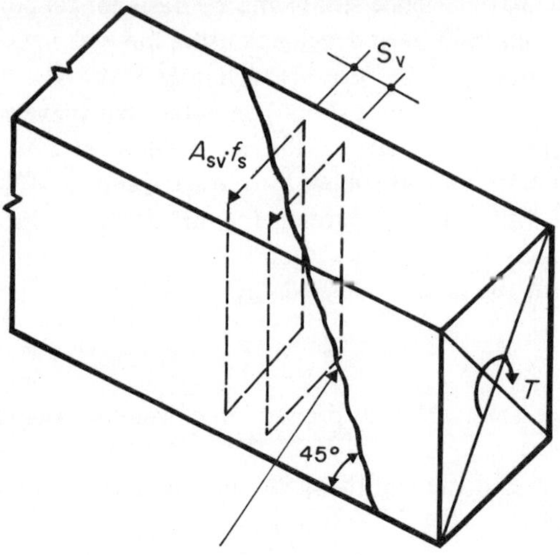

Fig. 1.10(a)

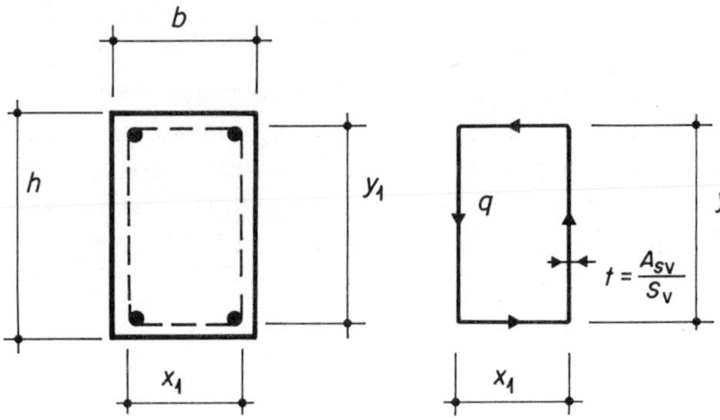

Fig. 1.10(b)

The stress in the links is given by

$$f_s = \frac{q}{t}$$

$$= q \frac{s_v}{A_{sv}}$$

and $$q = f_s \frac{A_{sv}}{s_v}$$

Thus $$T = 2 f_s \frac{A_{sv}}{s_v} x_1 y_1$$

or $$\frac{A_{sv}}{s_v} = \frac{T}{2 f_s x_1 y_1} \qquad \text{..... 1(xv)}$$

In equation 1(xv) it should be noted that A_{sv} is the area of one reinforcing bar. It is more usual to write the expression as

$$\frac{A_{sv}}{s_v} = \frac{T}{f_s x_1 y_1} \qquad \ldots\ldots 1(\text{xvi})$$

where A_{sv} is the area of both legs of the closed link. The same volume of longitudinal reinforcement must be provided in the distance Sv.

$$\text{thus} \qquad A_{s1} s_v = A_{sv}(x_1 + y_1)$$

$$\text{and} \qquad A_{s1} = A_{sv} \frac{(x_1 + y_1)}{s_v} \qquad \ldots\ldots 1(\text{xvii})$$

In equation 1(xvii) A_{s1} denotes the cross sectional area of the longitudinal bars and one longitudinal bar should be placed in each corner of the closed link. For a detailed treatment of reinforced concrete in torsion reference should be made to the excellent monograph by H. J. Cowan[4]. Although this work was published prior (1965) to CP 110, a very useful survey of building code requirements for torsion is given with guidance on permissible values for v_t and f_s, and in the final chapter by I. M. Lyalin, an ultimate strength approach is developed.

MEMBER ANALYSIS

As a starting point for the elastic analysis of indeterminate structural systems, it is necessary to carry out a preliminary analysis to obtain the relative stiffnesses

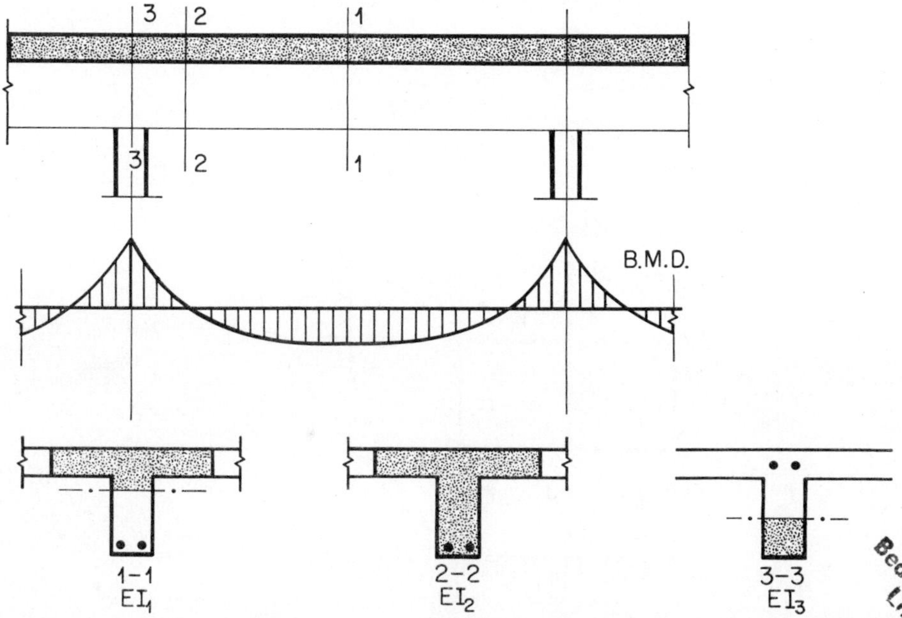

Fig. 1.11 An indication of the variation of flexural rigidity EI along the length of a continuous beam and slab system

of the members. For slabs and beams, trial thicknesses can be obtained from tables of basic span to depth ratios such as those given in CP 114 and CP 110. Approximate column sizes may be obtained by considering direct load only. The interaction of slabs with beams and the variability of flexural rigidity due to change from uncracked to cracked section along the length of a member make an accurate assessment of the true stiffness of the structure virtually impossible, and in the interests of simplicity, the entire concrete cross section making allowance for T-beam action is normally used for assessing relative stiffnesses. The use of the gross cross section or transformed section requires a prior estimate of the reinforcement area. An indication of the variation in flexural rigidity along the length of a continuous beam and slab system is shown in Figure 1.11, and to attempt to allow for this in the analysis of an indeterminate system is not realistic.

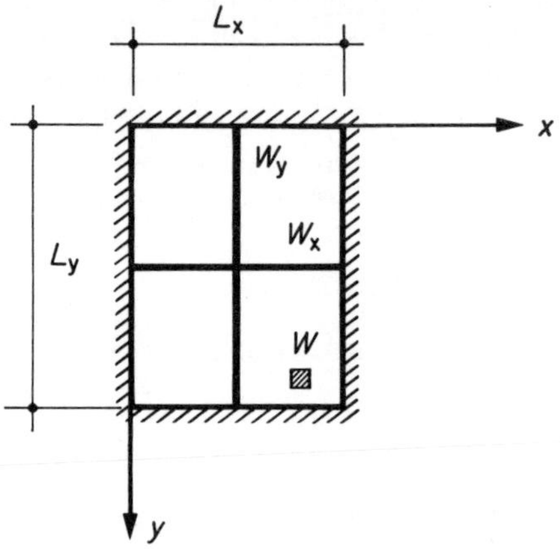

Fig. 1.12(a)

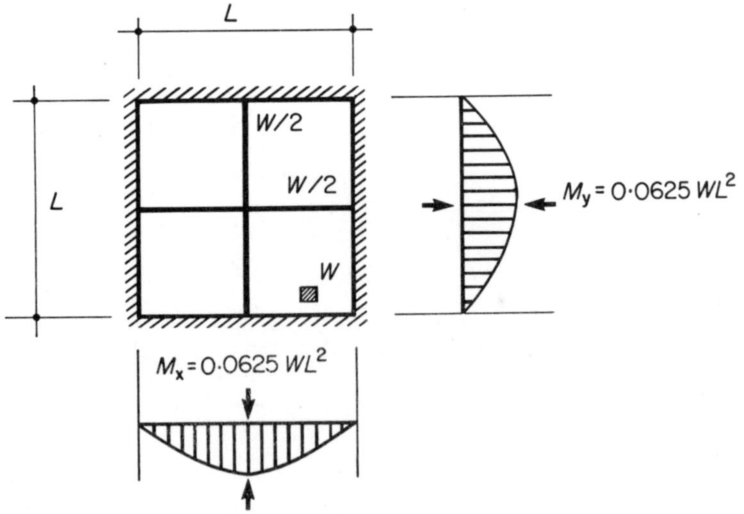

Fig. 1.12(b)

1 SLABS

Prior to World War II, when plastic theories were first used to evaluate the distribution of force actions on slabs, the Rankine-Grashof formula[5] was commonly employed in design offices, as illustrated in Figure 1.12(a)(b). The distributed loading is divided up inversely as the fourth power of the spans. For pure strip action (twisting ignored)

$$w = w_x + w_y \qquad \text{..... 1(xviii)}$$

and equating midspan deflections of central strips (Figure 1.12(a)),

$$\frac{5w_x L_x^4}{384EI} = \frac{5w_y L_y^4}{384EI}$$

$$\frac{w_x}{w_y} = \frac{L_y^4}{L_x^4} \qquad \text{..... 1(xix)}$$

Thus w_x and w_y may be evaluated. For the case of a square slab of side $L = L_x = L_y$

$$w_x = w_y = \frac{w}{2}$$

Thus the maximum bending moment is $0.0625wL^2$.

Values of w_x and w_y for various aspect ratios are given in Table 1.3

Table 1.3 Values of w_x and w_y for various aspect ratios; slab simply supported on all four sides

L_x	L_y	$\frac{L_y^4}{L_x^4}$	w_x	w_y
L	L	1	$0.5w$	$0.5w$
L	$1.25L$	2.44	$0.709w$	$0.291w$
L	$1.5L$	5.07	$0.835w$	$0.165w$
L	$1.75L$	9.38	$0.930w$	$0.097w$
L	$2.0L$	16.0	$0.941w$	$0.059w$

Using the (L_y^4/L_x^4) relationship for rectangular slabs, it can be seen from Table 1.3 that for an aspect ratio of 1.5 or greater the majority of the load is carried in the x direction.

A refinement of the above is to use the full plate equation which includes the contribution of twisting to the load distributing capacity of the slab, the equation of equilibrium being

$$\frac{d^2 M_x}{dx^2} + \frac{d^2 M_y}{dy^2} - \frac{2d^2 M_{xy}}{dx dy} = -w \qquad \text{..... 1(xx)}$$

The solution of this equation for other than simple configurations and loadings involves considerable labour, but fortunately there are some tabulated data available. Timoshenko[6] gives results for a number of loading cases and support conditions based on the solution of equation 1(xx), which are of great assistance in the design office. Where tabulated data are not available, the electronic computer has become of increasing help in the solution of more difficult problems. The reduction in bending moment when using the complete equation compared with strip action is about 25 per cent for the case of a simply supported plate of aspect ratio 1.0

$$M_x = M_y = 0.048\, wL^2$$

2 BEAMS

The elastic analysis of continuous beams may be carried out using standard techniques such as moment distribution or influence coefficients, but extensive tabulated data of the form illustrated in Figure 1.13 are available. The arrangement of dead and imposed loading should be that which causes the most severe force action. For continuous beams of approximately equal span the following arrangements are normally considered.

Condition for maximum support moment: two adjacent spans loaded with dead and imposed load, all other spans dead load.

Condition for maximum span moment: alternate spans loaded with dead and imposed load, all other spans dead load.

3 FRAMES

A multi-storey multi-bay framework subject to vertical and lateral loads is a highly indeterminate structure and the use of standard plane frame computer programs[7] would appear to be the obvious approach to the assessment of the distribution of force actions. However, there is a significant amount of labour involved in preparation of data and interpolation of the output. In the author's view, the most useful feature of a computer analysis is that joint translations are given in the output thus the designer can assess need to consider amplification of moments due to large sway deformations. A simple approach is to assume the column deflects to a sine curve, the properties of which are as follows (see Figure 1.14 for notation):

$$\text{area of half BMD} = N a_m \frac{L}{\pi}$$

$$\text{and} \quad \bar{x} = \frac{L}{\pi}$$

where N = direct load

$a_m = a_0 + a_1$

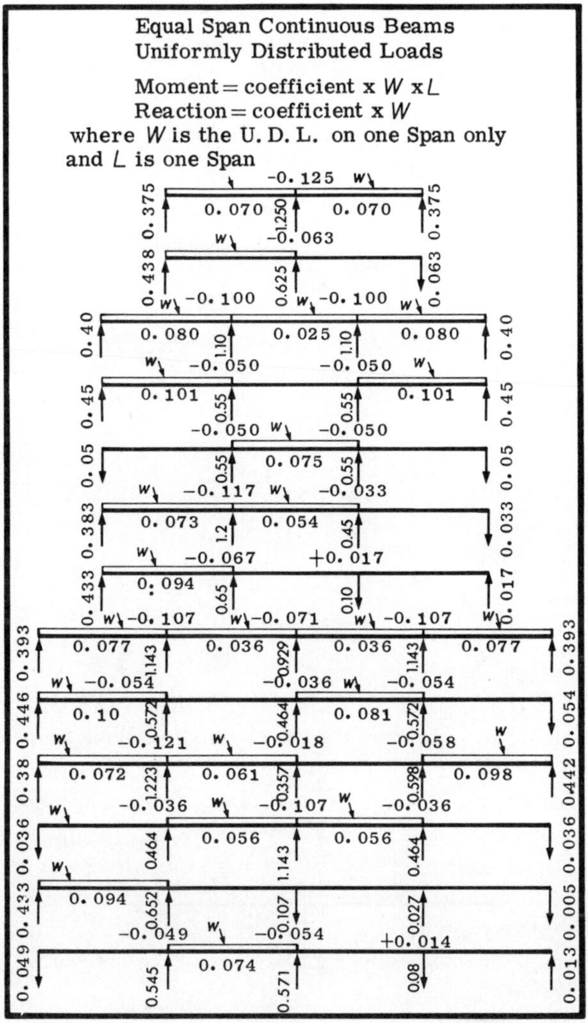

Fig. 1.13 Bending moment coefficients and reactions for continuous beams based on an elastic analysis, no redistribution

and a_0 = deflection due to primary applied bending moment M
a_1 = deflection due to secondary moment caused by load N

Using the moment area theorem

$$a_1 = \frac{A\bar{x}}{EI}$$

$$= Na_m \frac{L}{\pi}\frac{L}{\pi}\frac{1}{EI}$$

$$= Na_m \frac{L^2}{\pi^2 EI}$$

the Euler load $N_{crit} = \dfrac{\pi^2 EI}{L^2}$

thus $a_1 = \dfrac{N}{N_{crit}} a_m$

now $a_m = a_0 + a_1$

$$= a_0 + \frac{N}{N_{crit}} a_m$$

$$a_m \left(1 - \frac{N}{N_{crit}}\right) = a_0$$

$$a_m = \frac{a_0}{\left(1 - \dfrac{N}{N_{crit}}\right)}$$

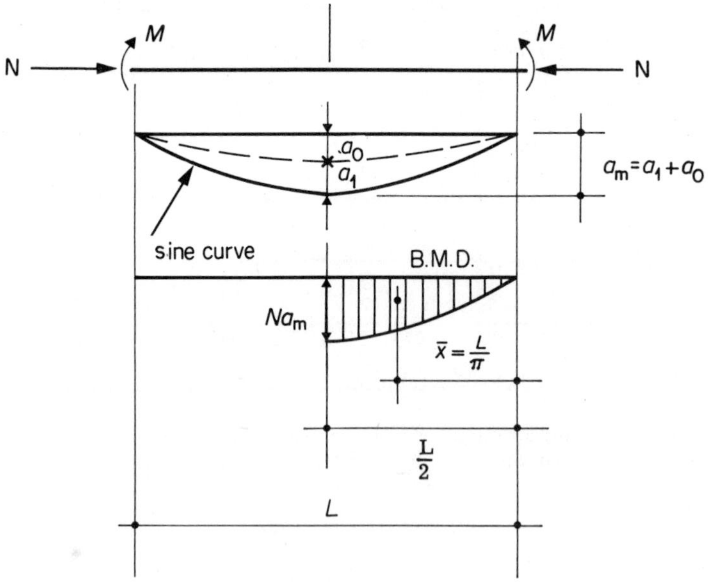

Fig. 1.14

Thus amplification of bending moment is

$$N_{am} = \frac{Na_0}{1 - \dfrac{N}{N_{crit}}} \qquad \text{..... 1(xxi)}$$

A detailed treatment of this problem with worked examples is given in Reference 8. The section is analysed for the direct load N plus the total moment $M + N_{am}$, where M is the primary applied moment. For the case of a cantilever column in Figure 1.15, $M = HL$, $a_0 = NL^3/3EI$ and $N_{crit} = \pi^2 EI/4L^2$. The beam/column problem is considered in detail in Chapter 3.

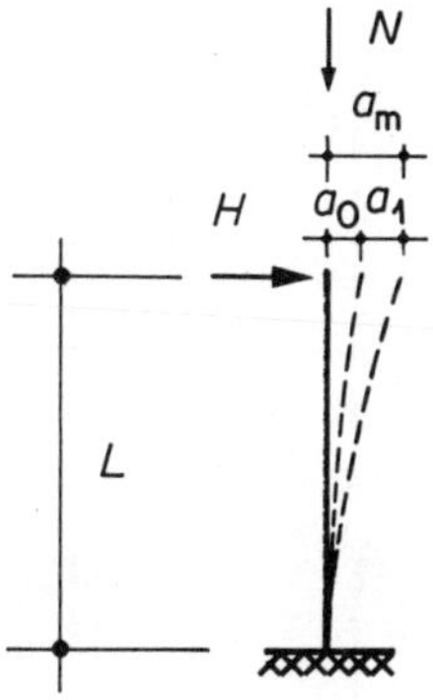

Fig. 1.15

In the absence of computer facilities, the following approach to the assessment of moments in frameworks subjected to vertical and lateral loads will, in general, give satisfactory results.

(*a*) *Lateral loads*

A statical analysis assuming points of contraflexure at mid column height, as in Figure 1.16.

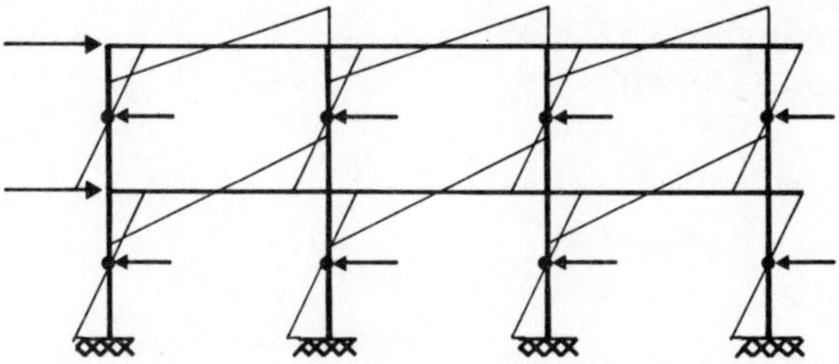

Fig. 1.16 Approximate analysis for lateral loads assuming point of contraflexure at mid column height

(b) *Vertical loads*
On a floor by floor basis using moment distribution or approximate bending moment coefficients as in Figure 1.17.

To summarize, the conventional approach to the design of reinforced concrete structures is based on the use of an elastic analysis to assess (a) the distribution of force actions within the structure, and (b) the strength of the section.

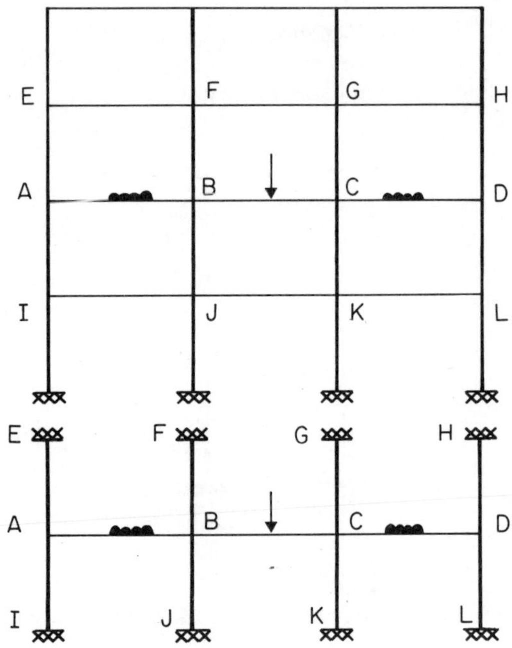

Fig. 1.17 Approximate analysis of floor A B C D for vertical loads. Assume fixity at E F G H and I J K L

It has long been established that reinforced concrete behaves in a manner intermediate between the elastic–plastic response of the reinforcement and the brittle behaviour of the concrete, which under normal conditions of loading is capable of very little plastic deformation. In the next chapter the development of ultimate strength theories leading to the comprehensive concept of multiple limit states will be surveyed.

2

The Development of Limit State Analysis

'The present customary means of calculation employ Hooke's law, a so called constant quantity (which however becomes a variable after a certain point), and the so called "permitted stresses" that represent the safety region before the ultimate strength of the concrete is reached, which is disproved by the fact that breaking does not generally occur on reaching the prescribed multiple of the "permitted stress". The present-day method of calculation is then inadequate and in addition undesirable in education, because it mediates false ideas of the true behaviour of a rectangular section under bending stresses. A new procedure must be based on its true behaviour, and must eliminate the illusory "constant quantity" and "permitted stresses". Simplicity is therefore more worth striving after than the greatest precision, which is naturally unobtainable; precision is impossible due to the rough estimate of the safety factor.'

R. Maillart (taken from *Schweizerische Bauzeitung*, January 1st 1938).

INTRODUCTION

Before introducing the concept of a limit state it is necessary to clarify the meaning of the word design. Structural design, a part of the overall design process, may be defined as the initial concept of the form and proportions of a structure required to serve one or more functions – for example, space enclosure to modify the environment or bridging a river. Structural analysis is a confirmation process, generally a calculation, which is carried out to satisfy the designer that:

(a) there is an adequate margin of safety against possible overload, defects in materials and inadequacies in analytical and construction procedures;
(b) the structure will remain serviceable during its useful life;
(c) there is a proper balance between safety and economy.

Unfortunately, the items listed above are difficult to quantify, and at present there are insufficient statistical data available to measure safety on a probabilistic basis. A vital consideration is the useful life of a structure and the longer the useful life the greater the probability of overload. Previous experience can give some information on the possible useful life of structures, and some typical examples are given:

road bridges	100–200 years
domestic structures	100 years
cathedrals	500 years
office blocks	50 years

However, sociological and technological progress is now so rapid that these figures may become completely unrealistic within the next decade. Thus it is necessary to adopt an analytical method which can readily be modified to take into account increasing knowledge of the properties of materials and the correlation of statistical data with regard to loading. This should be reflected in codes of practice, the format of which should be such that the designer is encouraged to make use of new developments in any particular field.

The use of reinforced concrete dates back to the end of the nineteenth century, and during the past 75 years the analysis of concrete structures has largely been based on the assumption that the composite material behaves in accordance with Hooke's law.

In most cases, concrete structures analysed on an elastic basis as outlined in Chapter 1, incorporating a stress factor of safety with limits imposed on span to depth ratios, have given adequate serviceability. However, it is worth noting that some of the finest examples of reinforced concrete work have been conceived by engineers such as Robert Maillart[9] (1872–1940) and Eugene Freyssinet[10] (1879–1962) who were very conscious of the inelastic behaviour of reinforced concrete under load. Over the past 35 years the recommendations of codes of practice

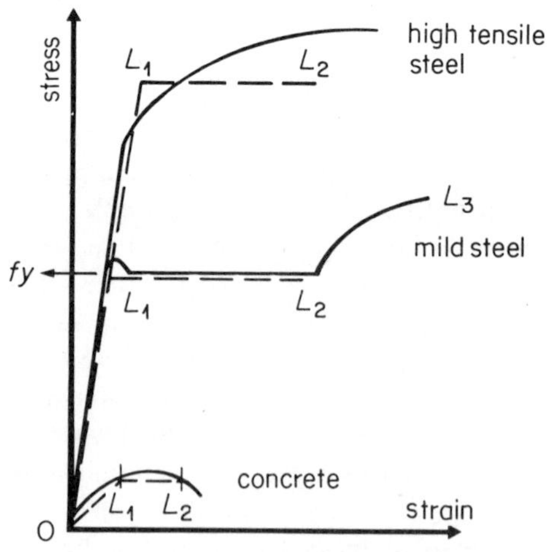

Fig. 2.1 Idealized stress/strain curves for steel and concrete

have gradually paid more attention to the inelastic behaviour of reinforced concrete. The response of a singly reinforced section to flexure gives a clear indication of the inelastic behaviour of the composite material. Figure 2.1 illustrates the stress/strain relationship for steel and concrete. The stress/strain curve for mild steel has a definite straight line portion OL_1 which represents the elastic range. At a stress f_y the steel yields and the portion $L_1 L_2$ represents the plastic range in which the steel undergoes extensive strain at constant stress. This is followed by a further increase in stress and strain (strain hardening) until fracture occurs at L_3. The dotted line represents a simplified form of this curve consisting of two straight lines OL_1 and $L_1 L_2$, which is sometimes referred to as the idealized stress/strain diagram. For high tensile steel there is not a well defined yield point but again the curve can be simplified to an idealized form $OL_1 L_2$. The stress/strain curve for concrete is approximately linear at low stress levels and again can be represented in an idealized form by the dotted line $OL_1 L_2$. The limiting concrete strain L_2 is generally in the range 0.003–0.004 and for the present argument a mean value of 0.0035 will be assumed. Yield of mild steel takes place at a strain value of about 0.002.

Referring to Figure 2.2 the various stages in the deformation of a concrete beam under flexure can now be examined taking due note of the stress/strain

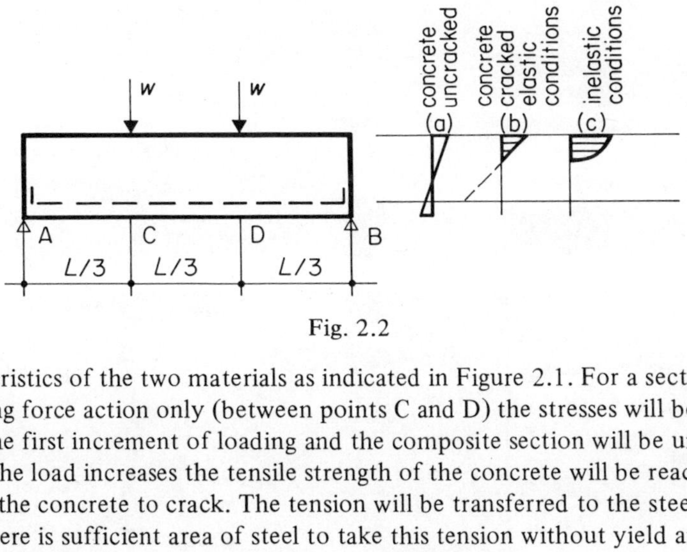

Fig. 2.2

characteristics of the two materials as indicated in Figure 2.1. For a section under a bending force action only (between points C and D) the stresses will be low under the first increment of loading and the composite section will be uncracked (a). As the load increases the tensile strength of the concrete will be reached causing the concrete to crack. The tension will be transferred to the steel provided there is sufficient area of steel to take this tension without yield and there is adequate bond between the steel and concrete.

The compressive stress will still be within the approximately elastic range and thus a straight line distribution may be assumed (b). With further increase in load the depth of the cracks will increase but the steel bonded to the concrete between the cracks will continue to take the tension. In the compression zone the extreme fibre stresses will no longer be within the approximately straight line portion of the stress/strain curve and thus the distribution of stress will be as in diagram (c).

Assuming adequate bond, the failure of the beam in flexure will depend on the percentage of reinforcement $A_s/bd = p$. If p is low the steel will reach its yield strain (approximately 0.002) before the concrete reaches its limiting strain of, say, 0.0035. With yield of the steel a moderate increment of load will cause large increase in crack depth and width, resulting in a gentle spalling of the concrete in the extreme compression fibres. If the value of p is high, the concrete will reach its limiting strain before the steel yields and the failure will be sudden and explosive. It is obviously desirable to proportion a reinforced concrete section to ensure that the steel yields before the concrete reaches its limiting strain thus giving ample warning of failure. A further advantage is that the plasticity of the section can be utilized when considering member analysis at ultimate conditions. Figure 2.3 illustrates the development of cracks in a reinforced concrete section reinforced to ensure the failure mode is gradual. This is known as an under-reinforced section; that is, the reinforcement percentage is such that the steel yields

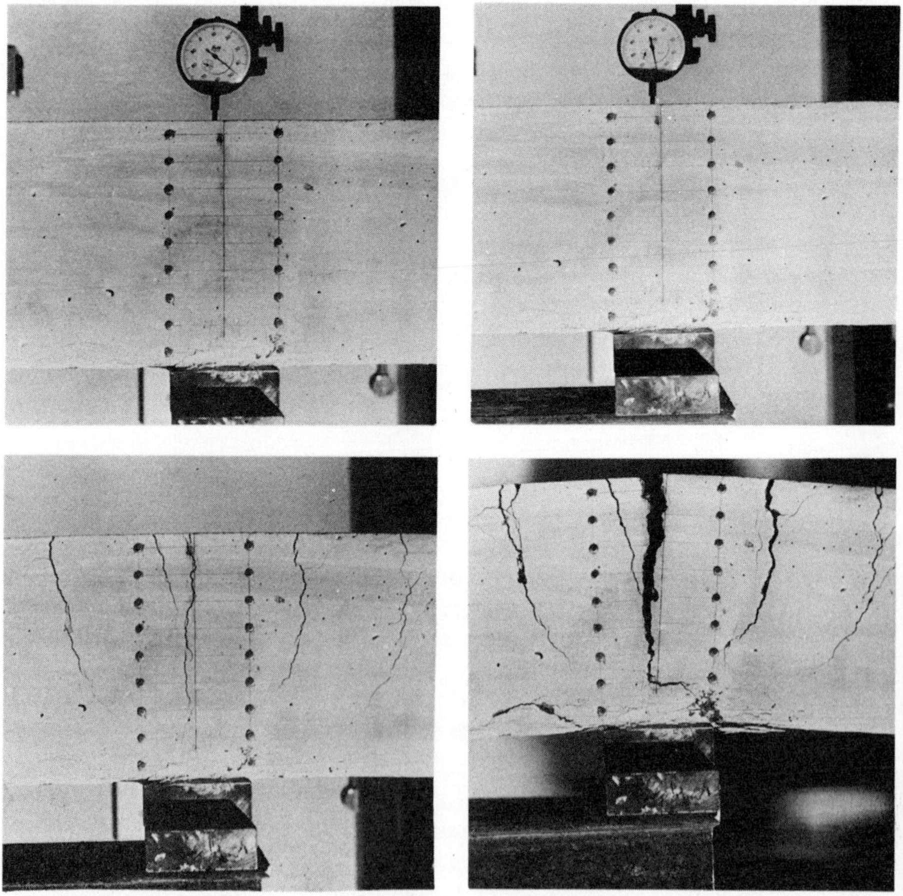

Fig. 2 (3) The development of cracks up to failure in an under-reinforced beam

before the concrete reaches its limiting strain. If the concrete reaches its limiting strain before the steel yields, the section is over-reinforced and characterized by sudden failure with little or no plasticity. Thus it would appear reasonable to base the analysis of a reinforced concrete section at ultimate conditions on inelastic behaviour. Further this could also apply to member analysis if the sections were lightly reinforced. Consider a two span continuous beam with uniformly distributed loading and reinforced to give equal resistance moments at the support and in the span. A conventional elastic analysis for a service load of w would give, for constant flexural rigidity, a bending moment at support B of $0.125\,wL^2$ (Figure 2.4). If the sections are under-reinforced the moment curvature relationship will be as idealized in Figure 2.5, which implies elastic

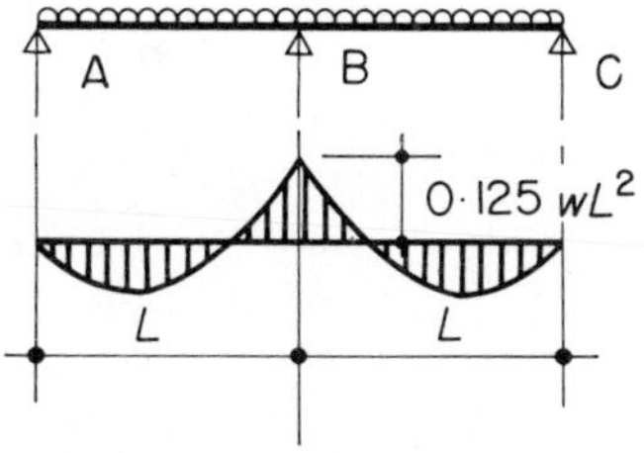

Fig. 2.4

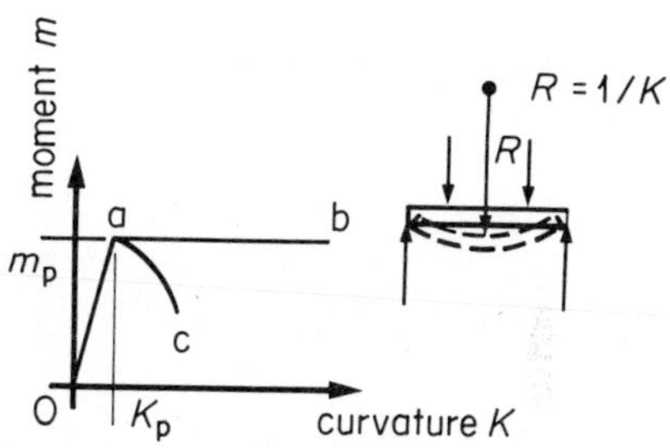

Fig. 2.5 Idealized moment/curvature relationship for a lightly reinforced section

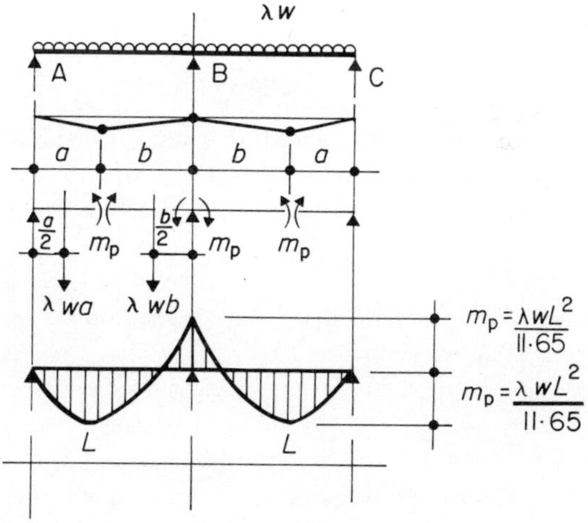

Fig. 2.6

behaviour up to yield of the steel and the moment m_p being maintained with increase in curvature beyond K_p. If the section is over-reinforced the moment curvature relationship would be as curve 0ac. Consider loading beyond the service value of w. The steel will yield first at support B and this section will behave in a plastic manner for further increase in load and will act as a plastic hinge. Thus with increase in load section B will continue to rotate at constant moment m_p until further hinges develop in the spans (Figure 2.6) resulting in the development of a mechanism at load λw. Considering the equilibrium of the system at load λw then

$$a + b = L \qquad \ldots\ldots 2(i)$$

$$m_p = \frac{\lambda w a^2}{2} \qquad \ldots\ldots 2(ii)$$

$$2m_p = \frac{\lambda w b^2}{2} \qquad \ldots\ldots 2(iii)$$

thus from equations 2(ii) and 2(iii)

$$\frac{1}{2} = \frac{a^2}{b^2}$$

$$b = \sqrt{2}a \qquad \ldots\ldots 2(iv)$$

Substituting equation 2(iv) in 2(i)

$$a + \sqrt{2}a = L$$

$$a = 0.414L$$

and from equation 2(ii) $$m_p = \frac{\lambda w (0.414L)^2}{2}$$

$$= \frac{\lambda w L^2}{11.65} \qquad \ldots\ldots 2(v)$$

Thus at ultimate conditions the moment distribution is completely different from that obtained from an elastic analysis. Hence it would appear that for the analysis of a reinforced concrete structure at ultimate conditions the following procedure could be adopted.

Member analysis: use inelastic analysis to determine the distribution of force actions.
Section analysis: strength of section assessed by inelastic analysis.

However, the above argument relates one force action only – bending – and consideration of other force actions produces complications which will be examined in Chapter 3. Also the analysis of a structure at ultimate conditions ignores the serviceability requirements. At the service load w the deformation of the structure and crack widths may be excessive. Hence it is necessary to consider both strength and serviceability, and for serviceability an elastic analysis is more appropriate.

It must be emphasized that an elastic analysis is still an essential part of the calculation process and it is undesirable to consider a single failure criterion. The introduction of multiple criteria for analysis in codes of practice for reinforced concrete structure will now be summarized.

During the 1930s substantial criticisms were formulated of the classical elastic method of assessing the strength of a reinforced concrete section. In a research programme carried out in Russia[11] emphasis was placed on both ultimate strength and unsuitability for service due to excessive deformation and cracking. As a result of this work, in 1938 the Russian Standards abandoned the elastic proportioning of reinforced concrete sections assuming a constant ratio α_e of the modulus of elasticity of steel to that of concrete. The form of the stress block adopted for flexure was as in Figure 2.7. Limitations were placed on the

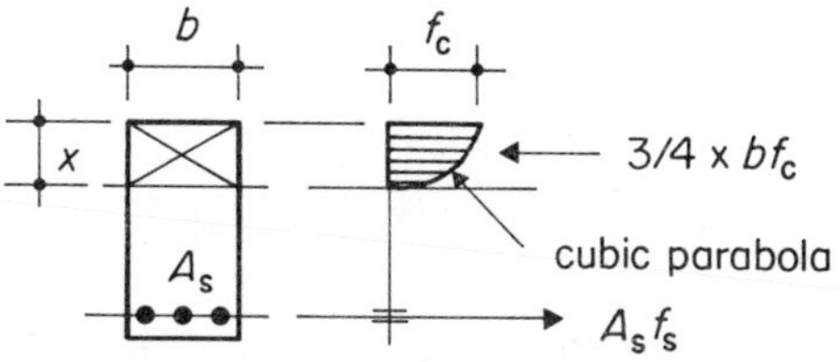

Fig. 2.7 Russian specification, section analysis for flexure (1938)

value of x to ensure sections were under-reinforced. In British codes of practice the elastic concept was maintained until 1957. The 1957 revision of CP 114, the British Standard Code of Practice for the Structural Use of Concrete introduced the so called 'load factor' approach to the analysis sections in flexure using a simplified rectangular stress block. The resistance moment equation is derived as follows. Assuming a parabolic stress block at ultimate conditions (Figure 2.8(a)) the average stress is two thirds of the maximum value. Due to

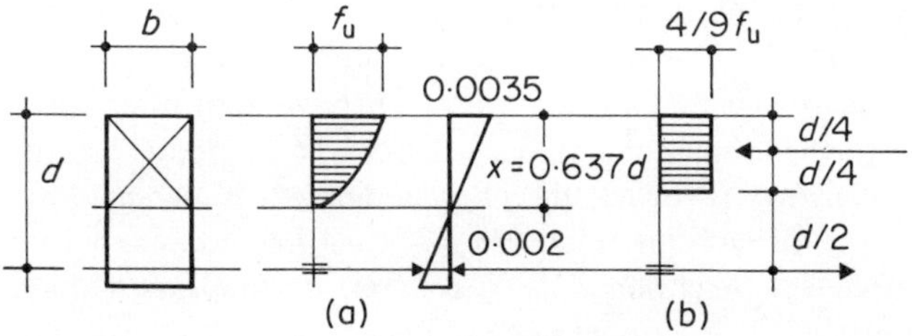

Fig. 2.8 CP114: 1957 stress block

the greater variability of concrete compared with steel the strength of the concrete is taken as two thirds of the maximum value and thus the average stress at ultimate conditions is taken as $\frac{2}{3} \times \frac{2}{3} = \frac{4}{9}$ the cube strength of the concrete. The value of x is limited to $0.5d$ to ensure an under-reinforced failure mode. Assuming the limiting concrete strain is 0.0035 and the yield strain for the steel is 0.002, then, for these values to occur simultaneously,

$$\frac{x}{d-x} = \frac{0.0035}{0.002}$$

thus $$x = 0.637d$$

Hence limiting the value of x to $0.5d$ ensures that the steel will yield before the concrete reaches its limiting strain. With $x = 0.5d$ the resistance moment of the section is given by (Figure 2.8(b))

$$M_u = \frac{4}{9} f_u b \frac{d}{2} \frac{3}{4} d$$

$$= \frac{f_u}{6} bd^2 \qquad \text{..... 2(vi)}$$

In CP 114 the resistance moment of the section is given by

$$M_w = \frac{p_{cb}}{4} bd^2 \qquad \text{..... 2(vii)}$$

where p_{cb} is the permissible compressive stress in the concrete at service load and is taken as $f_u/3$. Thus substituting $p_{cb} = f_u/3$ in equation 2(vii) gives

$$M_w = \frac{f_u}{12} bd^2 \qquad \text{..... 2(viii)}$$

Equations 2(vii) and 2(viii) give the same result, the former relating to ultimate load conditions and the latter to service load conditions with an overall or global load factor of $\lambda \doteq 2.0$.

$$= \frac{M_u}{M_w} = 2.0$$

Equation 2(vii) is thus of ultimate load format but disguised in terms of a working stress p_{cb}.

CP 114 : 1957 also allowed a 15 per cent adjustment of moments obtained by an elastic analysis which implies inelastic behaviour at ultimate conditions. Referring to the bending moment $M_B = 0.125\,wL^2$ in Figure 2.4 this value can be reduced to $0.106\,wL^2$ providing the span moment is increased by a corresponding amount. In 1958 the Institution of Civil Engineers appointed a committee to review existing knowledge of concrete up to ultimate load. The findings of this committee which worked in close liaison with the Comité Europeen Du Beton (CEB) were published[12] in 1962. In 1963 the CEB published draft recommendations for an International Code of Practice for Reinforced Concrete, and in 1970 the final document was produced.[13]

The CEB recommendations formed a guide to the preparation of CP 110 : Part 1 : November 1972, Code of Practice for the Structural Use of Concrete.[14] This code of practice may be used in place of CP 114, CP 115 and CP 116, which will still continue to exist. The CEB considered that the principles and recommendations should be based solely on experimental knowledge of the actual

Table 2.1 A summary of the development of Codes of Practice during the period 1948–1972

Date	*Deflection*	*Section analysis for flexure*
1948 CP 114	Cautionary note regarding deflection	Elastic theory, triangular stress block constant modular ratio $\alpha e = 15$
1957 CP 114	Table of limiting span to depth ratios	Load factor or elastic theory – use of simplified rectangular stress block incorporating permissible stress concept
1959 CP 115 The Structural Use of Prestressed Concrete		Dual criteria for analysis – ultimate strength and working stress – simplified rectangular stress block for ultimate strength
1963 ACI* (318–63) Building Code Requirements	Table of limiting span to depth ratios and procedure for calculating deflections	Choice of ultimate strength or working stress analysis – simplified rectangular stress block for ultimate strength
1963 CEB Draft Recommendations		Introduction of limit states – use of rectangular, parabolic or parabolic rectangular stress block for ultimate limit state
1965 CP 114 (revision)	Table of span to depth ratios with reduction factors for high steel and concrete stresses	As CP 114 : 1957
1970 CEB International Recommendations		Parabolic rectangular or simplified rectangular stress block for ultimate limit state
1971 ACI (318–71) Building Code Requirements	Table of limiting span to depth ratios and procedure for calculating deflections	Emphasis on ultimate strength design using load factors and capacity reduction factors with alternative design method using elastic theory
1972 CP 110	Table of span to depth ratios with modification factors and/or procedure for calculating deflections	Simplified rectangular or parabolic rectangular stress block for ultimate limit state

* ACI: American Concrete Institute

behaviour of the composite material conceived as forming a unified whole and that one of the most important tasks in concrete research is to further knowledge of the theory of structural safety. Safety is expressed in terms of the probability that the structure will not become unfit for its intended function during its useful life, that is, the structure will not reach a *limit state:*

> . . . 'The initial idea of referring to a single failure criterion has been replaced by the comprehensive concept of *limit states*. With this concept the local or the overall behaviour in all stages is considered: elastic, plastic, cracked, ultimate. In this approach to the problem a structure would have to be regarded as well designed if it could be shown that the probability of any limit state being attained is substantially constant for all the component members and for that structure as a whole and that, consequently, the latter possesses adequate and uniform structural safety' . . . (CEB pxv) 3.

The introduction of multiple criteria for design in codes of practice is summarized in Table 2.1. An important aspect of more recent codes of practice is that greater emphasis is being placed on controlling deformation and crack widths due to high steel stresses at service load implicit in an ultimate load analysis. This will be amplified in Chapter 4. In CP 110 the global or overall safety factor λ is replaced by partial safety factors – γ_m a reduction factor for values of mechanical strength and γ_f an enhancement factor for loading. The values of γ_f and γ_m are related to the limit state being considered. The terms characteristic strength and characteristic load are also introduced from which design values may be evaluated using the appropriate values of γ_f and γ_m.

Characteristic Strength

The characteristic strength of a material may be defined as the strength that is normally expected to be exceeded and may be defined statistically as follows:

$$f_k = f_m - k_1 \sigma$$

where f_k = characteristic strength
f_m = mean strength
σ = standard deviation
k_1 = factor that ensures that the probability of the characteristic strength not being exceeded is small

For concrete the value of k_1 is taken as 1.64 which ensures that not more than 5 per cent (1 in 20) of the test results will give a strength less than the characteristic value.

For steel the characteristic strength is at present taken as the minimum strength in the appropriate British Standard.

DESIGN STRENGTHS

The design strength f_D is obtained by dividing the characteristic strength f_k by the appropriate value of γ_m, that is

$$f_D = \frac{f_k}{\gamma_m} \qquad \ldots\ldots 2(ix)$$

In CP 110 the following values of γ_m are given.

(a) ULTIMATE LIMIT STATE

γ_m = 1.5 for concrete and 1.15 for steel.

(b) SERVICEABILITY LIMIT STATE

When assessing deflections the appropriate values of γ_m are 1.0 for both concrete and steel. When assessing stresses or crack widths at any section within a structure the values of γ_m should be taken as 1.3 for concrete and 1.0 for steel.

CHARACTERISTIC LOAD

The characteristic load may be defined as the load which is not likely to be exceeded during the useful life of the structure and can be expressed in the following manner.

$$F_k = F_m + k_2\sigma$$

where F_k = characteristic load
F_m = mean load
σ = standard deviation
k_2 = factor that ensures that the probability of the characteristic load being exceeded is small

At present there is insufficient statistical information available to enable characteristic loads to be defined in this way but loading surveys are being carried out by the Building Research Establishment.[15] Thus it is necessary to assume that the characteristic load is the load given in the BS Code of Practice for loading, CP 3: Chapter V: Parts 1 and 2, or other appropriate source.

DESIGN LOAD

The design load F_D is obtained by multiplying the characteristic load by the appropriate value of γ_f thus

$$F_D = \gamma_f F_k \qquad \ldots\ldots 2(x)$$

Considering dead and imposed loading only, CP 110 gives the following values:

(a) ULTIMATE LIMIT STATE

design dead load $= 1.4\,G_k$
design imposed load $= 1.6\,Q_k$

(b) SERVICEABILITY LIMIT STATE

design dead load $= 1.0\,G_k$
design imposed load $= 1.0\,Q_k$

where G_k = characteristic dead load
Q_k = characteristic imposed load

The numerical value preceding G_k and Q_k is the partial safety factor γ_f. It should be noted that the arrangement of loads should be that which causes the most severe force actions, and thus for the ultimate limit state $1.0\,G_k$ should be applied to the structure where more unfavourable conditions could occur. For further information on load combinations – dead + imposed + wind etc., reference should be made to clause 2.3.3 in CP 110.

The primary limit states covered in CP 110 are the ultimate limit state, and serviceability limit states relating to deflection and cracking. The normal procedure is to start with proportioning the structure to satisfy the requirements for the ultimate limit state and then consider the serviceability limit states. It is convenient to split the analysis into three separate sections for each limit state – loading, member analysis and section analysis. A broad outline of the procedure is given below and a detailed treatment of the ultimate and serviceability limit states is the subject of Chapters 3 and 4.

ULTIMATE LIMIT STATE

1 LOADING

As a starting point it is necessary to estimate the self weight of the structural system, and the designer will normally rely on approximate span to depth ratios, such as those in Tables 8 and 9 of CP 110, or rely on previous experience. The characteristic dead and imposed loads are multiplied by the appropriate partial safety factor γ_f, and the arrangement of loads on the system should be such as to cause the most severe force action. As an example, for continuous beams the following arrangements should be considered:

For maximum support moment: two adjacent spans loaded with $(1.4\,G_k + 1.6\,Q_k)$, all other spans $1.0\,G_k$.

For maximum span moment : alternate spans loaded with $(1.4\,G_k + 1.6\,Q_k)$, all other spans $1.0\,G_k$.

2 MEMBER ANALYSIS

This is carried out in order to determine the maximum values of the force actions, moment, shear, etc., due to the loadings derived in 1. For the analysis of slabs and beams there are a number of options available as below.

(a) *Slabs*

(i) An elastic analysis using plate theory, moment distribution, influence coefficients or other techniques with up to 30 per cent redistribution of moment depending on the neutral axis depth x.
(ii) Use of bending moment coefficients given in Tables 12 and 13 of CP 110, or other appropriate source.
(iii) Use of the yield line theory.
(iv) Use of the Hillerborg equilibrium method.

Options (i) and (ii) represent a conventional elastic approach to assessing the distribution of force actions at ultimate conditions with factored loads, but the provision for redistribution of moment implies inelastic behaviour. Options (iii) and (iv) are based on the concept of plastic behaviour of concrete slabs. The object of a yield line analysis is to postulate a fracture or yield line pattern from which the ultimate load can be determined by

(a) considering the equilibrium of the slab elements; or
(b) using the work equation.

The inherent difficulty of a yield line analysis is to determine the yield line pattern which will give the minimum collapse load. In general an upper bound solution is obtained (correct or unsafe). However, due to membrane effects, test results can show much higher failure loads than those predicted by the yield line theory. Thus the yield line theory may be used with confidence in the design office if a reasonable attempt is made to find the minimum failure load.

The Hillerborg equilibrium or strip method is a design approach in which the load is carried by a series of strips assumed to act like beams reinforced along their full length like beams carrying the same load. Thus it is possible to design a slab with variable reinforcement in contrast to the yield line theory which, in general, is used to analyse a slab with given uniform reinforcement. It is assumed that plasticity allows the deflections of perpendicular strips to become compatible.

(b) *Beams*

(i) An elastic analysis using moment distribution, influence coefficients or other techniques with up to 30 per cent redistribution of moment depending on the neutral axis depth x.

(ii) Use of approximate bending moment coefficients (see Table 4, CP 110, clause 3.3.4) for uniformly distributed loads with Q_K not greater than G_K.

(iii) Use of a plastic hinge analysis with restrictions on the neutral axis depth, or calculation of permissible and actual hinge rotations.

As with slabs, options (i) and (ii) represent a conventional elastic approach whereas option (iii) assumes that the continuous beam system can develop sufficient plasticity for hinges to form, which can rotate at substantially constant moment.

(c) *Frames*

The plastic hinge approach may also be applied to frameworks but there is a danger of elastic instability occurring prior to the development of the collapse mechanism. Thus extreme care must be exercised with this approach where the structural frame provides lateral stability. An alternative approach is to carry out an elastic analysis with factored loads using the procedure outlined in Chapter 1. Also redistribution in moment for structures over four storeys in height should be limited to 10 per cent.

In cases where lateral stability is provided by walls or bracing designed to resist all lateral forces it is reasonable to use a plastic approach to the design of the floor slabs and beams.

3 SECTION ANALYSIS

The starting point for the analysis of the section for the force actions obtained in Section 2 will be that of flexure. Use can be made of the design charts for a parabolic rectangular stress block, given in Part 2 of CP 110, or those for a simplified rectangular stress block, given in Chapter 5. An important consideration is the neutral axis depth x. For slabs, x will rarely exceed $0.3d$, and thus 30 per cent redistribution can be used as indicated in CP 110. However, with beams, x will often exceed $0.3d$, and, if 30 per cent redistribution has been assumed in the member analysis, strict compliance with the requirements of CP 110 will not be achieved. It is the author's view that adequate plasticity can be achieved with x up to $0.5d$ and this will be considered further in Chapter 3.

SERVICEABILITY LIMIT STATES

As with the ultimate limit state it is convenient to split the analysis into three sections for each serviceability limit state.

(a) DEFLECTION

For the deflection limit state the normal practice is to use basic span/effective depth ratios with modification factors depending on the service stress and the amounts of tension and compression reinforcement which influence the stiffness of the section. An alternative is to estimate the deflection by means of a calculation and the procedure is outlined below:

1 *Loading*

The value of γ_f will, in general, be unity and thus the characteristic dead and imposed loads will be appropriate for the member analysis. The arrangement of loads should be such as to cause the most severe deformation. As an example if maximum deflection in the end span of a system of continuous beams is to be calculated then the arrangement of loading would be: end span $1.0\,G_k + 1.0\,Q_k$, penultimate span $1.0\,G_k$ and the arrangement repeated for subsequent spans.

2 *Member analysis*

The member analysis is carried out on an elastic basis with the assumption of constant flexural rigidity between supports, and use can be made of bending moment coefficients such as those in Figure 1.13.

For more complex loading systems and variable spans standard analytical techniques such as moment distribution or influence coefficients may be adopted.

3 *Section analysis*

The deflection at any point along the length of a beam may be expressed in the form

$$a = k_1 \frac{wL^3}{EI} \qquad \ldots\ldots 2(\text{xi})$$

where w is the total load
L is the span
EI the flexural rigidity
k_1 a coefficient depending on the type of loading and support conditions

For design purposes it is sufficiently accurate to assume that the midspan deflection approximates to the maximum value and using the relationship $M = k_2 wL$, where M is the midspan bending moment, and equation 2(xi) can be written as

$$a = \frac{Mk_1L^3}{k_2LEI}$$

$$= kL^2 \frac{M}{EI}$$

$$= kL^2 \frac{1}{r_b} \qquad \ldots\ldots 2(\text{xii})$$

where $1/r_b$ is the curvature at midspan (for cantilevers at the support section).

Equation 2(xii) forms the basis for the simplified method of calculating deflections given in CP 110. An important consideration is the evaluation of $1/r_b$ which in turn is related to the flexural rigidity *EI*. In order to obtain a realistic assessment of the deflection it is necessary to allow for the stiffening effect of the concrete in the tension zone. Use of equation 1(viii) will give an overestimate of the deflection and the procedure for including the stiffening effect of the concrete in the tension zone is given in Chapter 4. The influence of creep on the curvature $1/r_b$ is also covered.

The final deflection of a reinforced concrete member will be influenced by a number of factors, the proportion of the total load that is permanent, the amount of tensile and compressive reinforcement, creep and shrinkage, and calculations may only be of an approximate nature. The allowable total deflection will depend on the function of the structure, but as a guide it is to be noted that the table of basic span/effective depth ratios given in CP 110 (Table 8) should only be used for spans greater than 10 m if a deflection of span/250 is considered to be acceptable.

(b) CRACKING

The durability of a concrete structure will depend primarily on detailing the reinforcement to ensure that crack widths are limited to values which prevent resistance to moisture penetration and thus corrosion of the reinforcement. The surface width of cracks should, in general, not exceed 0.3 mm in an internal non-aggressive environment, but in other situations it may be necessary to reduce this value to 0.1 mm. Cracking is a variable phenomenon and it is not possible to be specific about crack widths nor is it possible to calculate them with a great degree of accuracy due to the number of interactions involved. The CP 110 formula for assessing crack widths is based on the correlation of a large number of measurements, the crack width primarily depending on

(i) the proximity to the point considered of reinforcing bars perpendicular to the cracks;
(ii) the proximity of the neutral axis to the point considered;
(iii) the average surface strain at the point considered.

The development and application of this equation is considered in Chapters 4 and 5 and an outline of the general procedure dealing with the limit state of excessive cracking is given:

1 *Loading*

The arrangement of loads will be in accordance with that for the ultimate limit state as the maximum flexural crack widths will be associated with maximum support and span moments. The value of γ_f will in general be unity.

2 *Member analysis*

The member analysis is carried out on a normal elastic basis in the same manner as for the limit state of excessive deformation.

3 *Section analysis*

The analysis of the section is carried out to determine the strain in the steel and then by proportion the strain at the level at which the crack width is to be measured. At any section the bending moment is known from the elastic analysis and for an appropriate value of the modular αe and known value of the reinforcement percentage from the ultimate limit state analysis, equation 1(ii) or Table 1.1 may be used to determine the neutral axis factor x and thus the lever arm z.

The steel stress is obtained from the relationship $f_s = M/A_s z$ and hence for a known value of the modulus of elasticity, the steel strain may be evaluated. However, this value is subsequently modified to allow for the stiffening effect of the concrete in the tensile zone.

To summarize, for the ultimate limit state it is now possible to use an inelastic analysis to determine the distribution of force actions in a structure and to assess the strength of sections. However, the distribution of moments may differ considerably from those obtained from an elastic analysis and thus an analysis for serviceability may be required, with particular reference to cracking and deformation. The calculation procedure will initially appear to be somewhat tedious, but with experience it will be possible to omit a number of stages; but the designer should not lose sight of the basic object of a limit state approach to analysis, that is, the structure is considered in terms of *strength* and *serviceability*.

3

Ultimate Limit States

This chapter is primarily concerned with the options listed in Chapter 2 which relate to the inelastic analysis of slabs and beams at ultimate conditions under the general headings of member and section analysis. The treatment of section analysis will adhere closely to the clauses in CP 110 relating to flexure, shear, torsion and bond. Columns will be treated separately with emphasis on the development of the CP 110 expression for amplification of moment.

MEMBER ANALYSIS

(a) YIELD LINE THEORY

The first reference to an inelastic approach for the determination of the distribution of force actions in a structure, to appear in a British Code of Practice, was in CP 114 : 1957: 'The ultimate bending moments to be allowed for should be deduced from analysis in which the load is 1.8 times the working (dead + imposed) and due regard is given to redistribution of moments that would occur before failure of the slab or beam by the use of Johansen's yield line theory or other acceptable method.' (CP 114 : 1957, clause 314c).

The origins of the yield line theory date back some 50 years, and early investigators such as Bach, Graf and Ingerslev derived solutions for the collapse mode of rectangular slabs from considerations of equilibrium provided by moments alone – shear and twisting force actions were not included. A more rigorous treatment was pioneered by Johansen in 1943 and since then the theory has been the subject of a detailed investigation by a number of research workers; for an in depth treatment of its development and application, reference should be made to three stimulating texts[16–18] by L. L. Jones and R. H. Wood. The following is a brief survey of the background to the theory and its application to various slab configurations.

As indicated in Chapter 2 the object of a yield line analysis is to postulate a yield line pattern from which the ultimate load of the slab can be evaluated for a given arrangement of reinforcement. It has been established that a lightly reinforced section has a moment/curvature relationship which can be idealized to a bi-linear form (Figure 2.5). Thus it may be assumed that the plastic deformation is large compared with the elastic deformation. Consider a square slab with simply supported edges, as in Figure 3.1(a), subjected to uniformly distributed loading reinforced with a square mesh of equal size bars placed in the bottom (isotropic reinforcement). This will give equal resistance moments in orthogonal

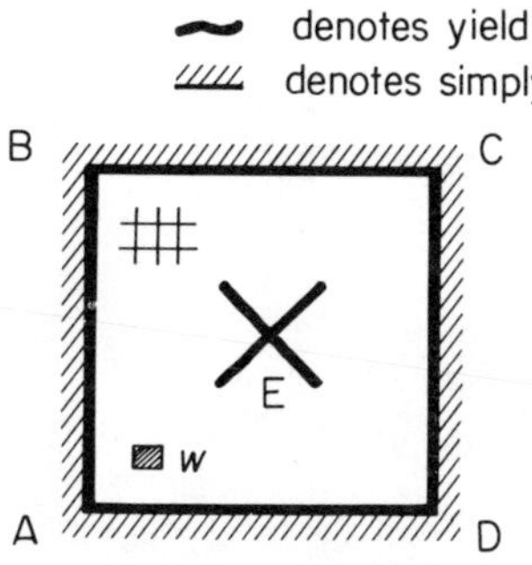

Fig. 3.1(a)

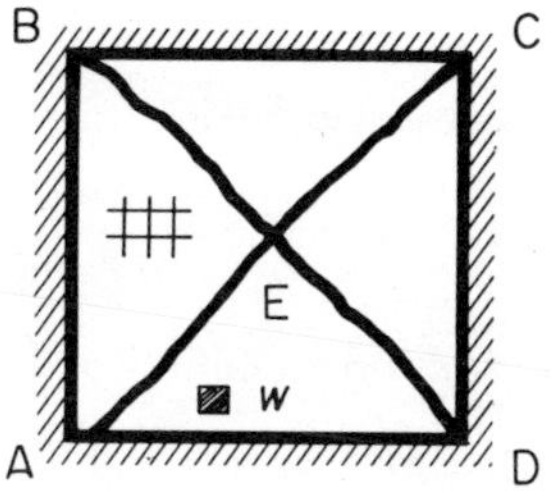

Fig. 3.1(b)

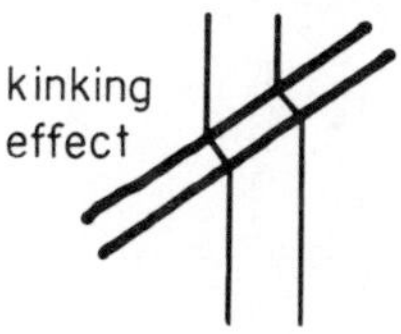

Fig. 3.1(c)

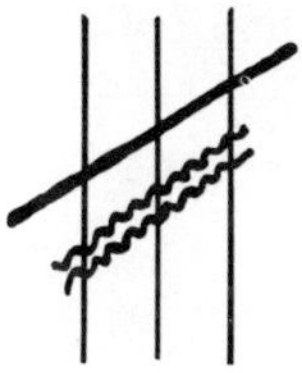

Fig. 3.1(d)

directions – the slight difference in effective depth for the two layers of reinforcement being ignored. At service load the response of the slab will be elastic, the maximum steel stress occurring at E. As the load is increased, the steel will reach its yield stress at E and with further increase in load the moment of resistance will not increase significantly (Figure 2.5), and thus sections adjacent to E will also yield and the yield lines will eventually extend to the slab corners (Figure 3.1(b)). The yield line pattern is now such that the slab is a mechanism and if it can be assumed that the ultimate resistance moment is achieved along the yield lines, then the slab can be analysed and its ultimate load evaluated. Johansen assumed that the reinforcing bars do not kink along a yield line, that is, they develop plastic behaviour in their original direction (Figure 3.1(c)). Some authors do not accept Johansen's assumption[19] but the 'kinking effect' is outside the scope of this simplified treatment. Using Johansen's assumption each yield line may be considered as an infinite succession of straight lines disposed stepwise,

Figure 3.1(d). It is now necessary to consider the equilibrium of the bending and twisting moments which act along a yield line. For a slab reinforced in one direction only, Figure 3.2, where m is the ultimate moment per unit length and the

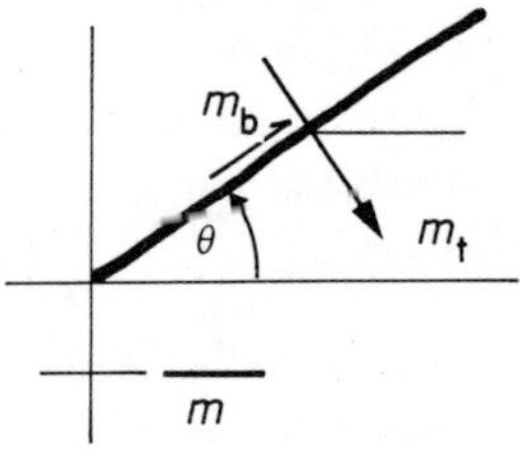

Fig. 3.2

yield line makes an angle θ with this direction, then resolution in the direction of m gives

$$m = m_b + m_t \tan\theta \qquad \text{.....} \; 3(i)$$

where m_b and m_t are the bending and twisting moments per unit length along the yield line. If moments are resolved at right angles to m

$$m_b = m_t \cot\theta \qquad \text{.....} \; 3(ii)$$

From equations 3(i) and 3(ii),

$$m = m_t(\cot\theta + \tan\theta)$$

$$= m_t\left[\frac{\cos\theta}{\sin\theta} + \frac{\sin\theta}{\cos\theta}\right]$$

$$= m_t\left[\frac{\cos^2\theta + \sin^2\theta}{\sin\theta\cos\theta}\right]$$

$$\text{and } 1 = \cos^2\theta + \sin^2\theta$$

$$\text{thus } m_t = m\sin\theta\cos\theta \qquad \text{.....} \; 3(iii)$$

From equations 3(i) and 3(ii),

$$m = m_b + \frac{m_b}{\cot\theta} . \tan\theta$$

$$= m_b\left[1 + \frac{\sin\theta}{\cos\theta} \cdot \frac{\sin\theta}{\cos\theta}\right]$$

$$= m_b\left[1 + \frac{\sin^2\theta}{\cos^2\theta}\right]$$

$$= m_b\left[\frac{\cos^2\theta + \sin^2\theta}{\cos^2\theta}\right]$$

$$\text{thus } m_b = m\cos^2\theta \qquad \text{.....} \; 3(iv)$$

If the slab is reinforced in more than one direction, then the separate effects may be added

$$m_b = \sum_{i=1}^{n} m_i \cos^2\theta$$

$$m_t = \sum_{i=1}^{n} m_i \sin\theta \cos\theta$$

For a slab reinforced in two directions at right angles m and μm, see Figure 3.3,

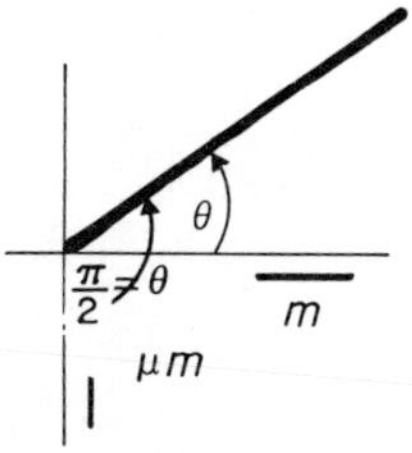

Fig. 3.3

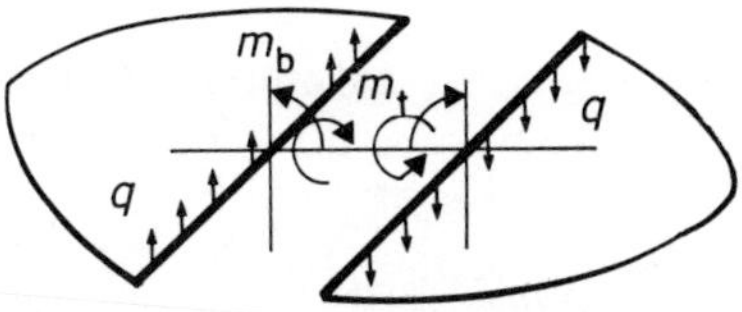

Fig. 3.4

$$m_b = m\cos^2\theta + \mu m \cos^2\left(\frac{\pi}{2} + \theta\right)$$

$$= m(\cos^2\theta + \mu \sin^2\theta)$$

and for $\mu = 1$ (isotropic reinforcement),

$$m_b = m \qquad \ldots\ldots 3(\text{v})$$

$$\text{and } m_t = m \sin\theta \cos\theta + \mu m \sin\left(\frac{\pi}{2} + \theta\right) \cos\left(\frac{\pi}{2} + \theta\right)$$

$$= m(1 - \mu)\sin\theta\cos\theta$$

thus for $\mu = 1$, $m_t = 0$ ..... 3(vi)

Thus for an isotropically reinforced slab, $m = m_b$ and the twisting moment is zero. Returning to Figure 3.1(a) the slab may be analysed by

(a) considerations of equilibrium;
(b) the work equation.

Using method (a) the forces and moments acting along a yield line are, see Figure 3.4:

(i) bending moments m_b;
(ii) twisting moments m_t;
(iii) shear forces q;

Johansen replaced[16] the twisting moments and shear forces by nodal forces Q for which he established a number of theorems, one of which being that at the junction of yield lines governed by the same mesh each of the nodal forces is zero. In Figure 3.5(a) element ABE which may be considered as a rigid element (elastic deformations small) it must be in equilibrium under the external loading w, the ultimate moments m, the nodal forces Q and the support reactions. In this case $m_b = m$ and $Q = 0$ thus

$$mL = \frac{wL^2}{4}\,\frac{L}{6} \quad \text{(moments about AB)}$$

$$m = \frac{wL^2}{24} \qquad \ldots\ldots 3(\text{vii})$$

Unfortunately, therc are a number of 'breakdown cases' with regard to the use of Johansen's nodal force theorems[18] and an alternative approach, method (b), can be used.

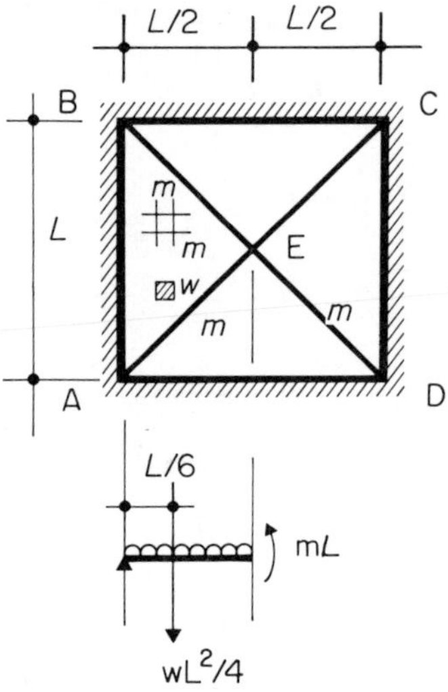

Fig. 3.5

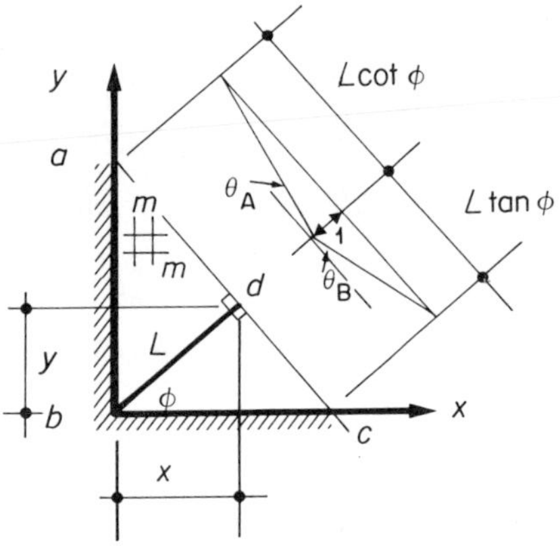

Fig. 3.6

Using the equality of internal and external work, m_t and q along a yield line undergo equal and opposite amounts of work and thus can be eliminated from the work equation. Further it is convenient to consider components of work. From Figure 3.6 for unit displacement at d, the internal work (moment x rotation) is given by

$$\Sigma M\theta = mL(\theta_A + \theta_B)$$

$$= mL\left[\frac{1}{L\cot\phi} + \frac{1}{L\tan\phi}\right]$$

$$= m(\tan\phi + \cot\phi)$$

Now for $m = m_x = m_y$, $x = L \cos \phi$ and $y = L \sin \phi$

$$\Sigma(m_x \theta_x x + m_y \theta_y y) = m \left[L \cos \phi \frac{1}{L \sin \phi} + L \sin \phi \frac{1}{L \cos \phi} \right]$$

$$= m(\tan \phi + \cot \phi) = \Sigma M\theta$$

Thus the components of work can be expressed as moment per unit length along an axis of rotation x length of projection x rotation about the axis under consideration. For Figure 3.5, considering the rigid element ABE, with unit displacement at E,

$$\text{the external work} = \frac{wL^2}{4} \times \frac{1}{3}$$

$$\text{the internal work} = m \times L \times \frac{2}{L} \text{ (rotation is about } x\text{-axis only)}$$

Thus for the four elements

$$4 \times \frac{wL^2}{4} \times \frac{1}{3} = m \times L \times \frac{2}{L} \times 4$$

$$m = \frac{wL^2}{24}$$

This result is identical to equation 3(vii).

It must be emphasized that the yield pattern shown in Figure 3.5(b) is merely a postulation and it is necessary to examine alternative patterns which may result in a lower value of w than $24m/L^2$. It is possible for a yield line to fork before it reaches the corner of a slab, forming what is known as a corner lever, Figure 3.7. Yield line patterns which have corner levers are more critical, but the error

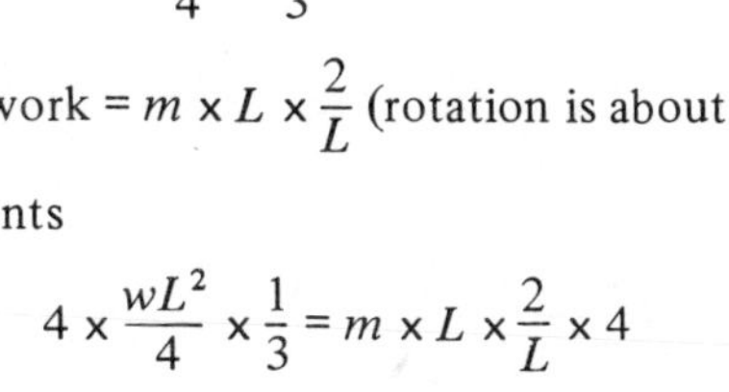

Fig. 3.7 Yield lines forking at the corners of a rectangular slab

involved for square slabs is small and is related to the ratio of top to bottom reinforcement $\mu m/m = \mu$. The following results apply to a square slab with uniformly distributed loading:[20]

μ	m
0	$wL^2/22$
0.25	$wL^2/23$
0.5	$wL^2/23.6$
1.0	$wL^2/24$

In many cases the general form of the work equation can be expressed as

$$m = \frac{u}{v} \text{ where } u \text{ and } v \text{ are functions of } x$$

thus
$$\frac{dm}{dx} = \frac{v\dfrac{du}{dx} - u\dfrac{dv}{dx}}{v^2}$$

for a maximum value of m

$$\frac{dm}{dx} = 0$$

and
$$0 = v\frac{du}{dx} - u\frac{dv}{dx}$$

hence
$$\frac{u}{v} = \frac{\dfrac{du}{dx}}{\dfrac{dv}{dx}} \qquad \ldots\ldots 3(\text{viii})$$

The application of this result will be illustrated in Example 1. The following examples illustrate the application of the work equation to a number of slab configurations and loadings. Extensive tabulated results are given in Reference 20 and there are a number of numerical examples in Chapter 5.

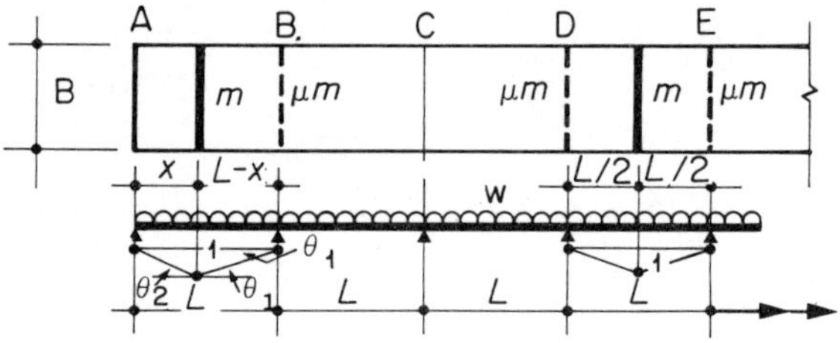

Fig. 3.8 Continuous slab with U.D.L. of w

Example 1, Figure 3.8

In this example consider straight yield lines developing in the end span AB under a uniformly distributed loading w

$$\text{external work} = wBx\frac{1}{2} + wB\frac{(L-x)}{2}$$

$$= \frac{wB}{2}(x + L - x)$$

$$= \frac{wBL}{2}$$

$$\text{internal work} = \mu mB\theta_1 + mB(\theta_1 + \theta_2)$$

$$= mB(\mu\theta_1 + \theta_1 + \theta_2)$$

$$= mB\left(\mu\frac{1}{L-x} + \frac{1}{L-x} + \frac{1}{x}\right)$$

Equating internal and external work,

$$\frac{wL}{2} = m\left[\frac{1}{L-x}(1+\mu) + \frac{1}{x}\right]$$

$$m = \frac{wL}{2}\,\frac{1}{\left[\dfrac{1+\mu}{L-x} + \dfrac{1}{x}\right]}$$

$$= \frac{wL}{2}\left[\frac{x(L-x)}{x(1+\mu) + (L-x)}\right] = \frac{u}{v}$$

For a maximum, from equation 3(viii),

$$\frac{x(L-x)}{x(1+\mu) + (L-x)} = \frac{L-2x}{\mu}$$

which simplifies to

$$L^2 - x^2\mu - 2xL = 0 \quad \text{and putting } \mu = 1$$

$$x = L\left[\frac{\sqrt{8}}{2} - 1\right]$$

$$= 0.414L$$

$$\text{and} \quad m = \frac{wL}{2}\left[\frac{1}{\dfrac{2}{L-0.414L} + \dfrac{1}{0.414L}}\right]$$

$$= \frac{wL^2}{2}\,\frac{1}{5.825} = \frac{wL^2}{11.65} \qquad \text{..... 3(ix)}$$

This result is identical to equation 2(v).

For an internal span DE,

$$\text{external work} = \frac{wBL}{2}$$

$$\text{internal work} = m \times B \times \frac{2}{L} \times 2 + \mu m \times \frac{2}{L} \times 2 \times B$$

$$\text{and for} \quad \mu = 1 = \frac{8Bm}{L}$$

$$\text{thus} \qquad m = \frac{wL^2}{16} \qquad \text{..... 3(x)}$$

Thus for a one way continuous slab with straight yield lines under uniformly distributed loading, design for the ultimate limit state could be based on equations 3(ix) and 3(x) with the cautionary note that the moment distribution at service load will be completely different. Thus a serviceability check is required, see Example 5.02. If it is assumed that the yield line in the end span occurs at $x = L/2$, then

$$\text{internal work} = m \times B \times \frac{2}{L} \times 2 + \mu m \times B \times \frac{2}{L}$$

$$\text{and for } \mu = 1 = \frac{6mB}{L}$$

$$\text{external work} = \frac{wBL}{2}$$

$$\text{thus } m = \frac{wL^2}{12} \qquad \ldots\ldots 3(\text{xi})$$

Thus a slab designed for this moment will give an unsafe result (in this case the error is very small) as the value of m will be exceeded at another section. In general, postulation of a yield pattern will give the correct or too high a collapse load – an upper bound solution. However, providing a reasonable attempt has been made at predicting the critical pattern, the error involved will be small and there is, in many cases, an enhancement of load carrying capacity due to membrane effects which are considered in the next section.

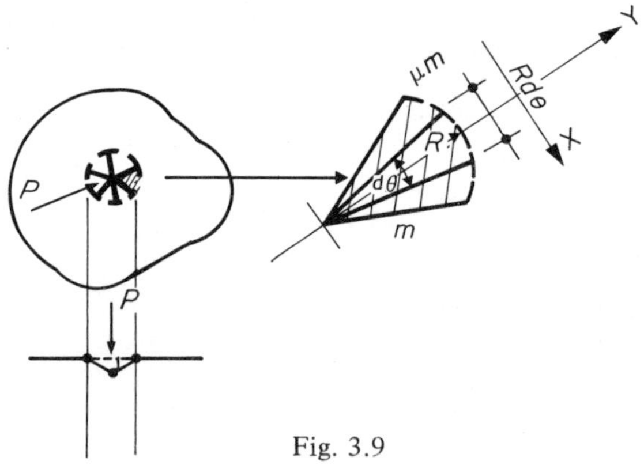

Fig. 3.9

Example 2, Figure 3.9
A point load P acts on an isotropically reinforced slab which is assumed to extend for an infinite distance in all directions. A possible failure mode is a fan type pattern, the deflected form being an inverted cone. Using axes of rotation X and Y and considering an element of the fan, the components of work are

$$= m_x\theta_x L_x + m_y\theta_y L_y$$

There is no rotation about the X axis thus the total internal work for the element is $m_x\theta_x L_x$ and with unit displacement under P

$$\text{internal work} = (m + \mu m) \times R\text{d}\theta \times \frac{1}{R}$$

$$= m(1 + \mu)\text{d}\theta \qquad \ldots\ldots 3\text{(xii)}$$

$$\text{total internal work for fan complete} = 2\pi(m + \mu m)$$

$$\text{external work} = P \times 1$$

Thus for equality of work done,

$$P = (m + \mu m)2\pi$$

$$\text{and if} \quad \mu = 1 \; P = 4\pi m$$

$$m = \frac{P}{4\pi} \qquad \ldots\ldots 3\text{(xiii)}$$

Equation 3(xii) is a useful result and can be used as a starting point for investigating the effect of point loads on bridge decks or examining possible failure modes for flat slab floors.

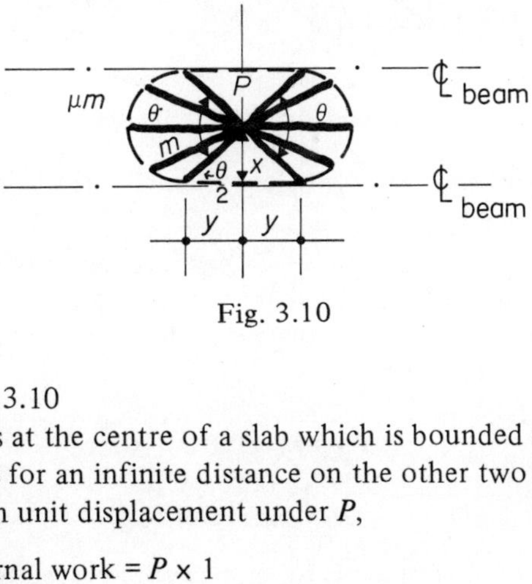

Fig. 3.10

Example 3, Figure 3.10
A point load P acts at the centre of a slab which is bounded on two sides by stiff beams and extends for an infinite distance on the other two sides. For the yield pattern shown with unit displacement under P,

$$\text{external work} = P \times 1$$

$$\text{internal work} = (m + \mu m)2\theta + (m + \mu m)2y \times \frac{1}{x} \times 2$$

$$\text{thus} \qquad P = (m + \mu m)2\theta + (m + \mu m) \times 4 \times \cot\frac{\theta}{2}$$

$$\text{for a minimum,} \qquad \frac{\text{d}P}{\text{d}\theta} = 0$$

$$0 = 2(m + \mu m) - 2(m + \mu m)\ \text{cosec}^2\ \frac{\theta}{2}$$

$$\text{cosec}^2\ \frac{\theta}{2} = 1$$

$$\text{cosec}\ \frac{\theta}{2} = 1$$

$$\text{giving}\quad \frac{\theta}{2} = \frac{\pi}{2}$$

$$\text{thus}\quad \theta = \pi \quad \text{i.e. full fan}$$

$$\text{and putting}\quad \mu = 1 \quad P = 4\pi m$$

Example 4, Figure 3.11

A point load P acts at (a) the corner, and (b) some distance along a cantilever slab.

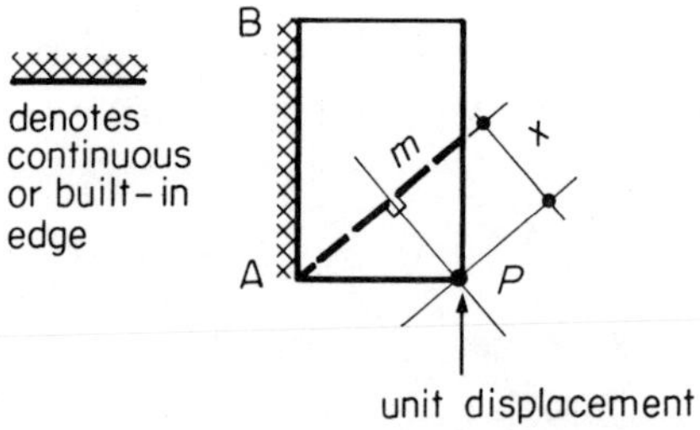

Fig. 3.11 Note collapse of whole balcony could occur by development of a yield line along AB

Case (a). One possible mode is for the slab to break off at the corner, and thus for isotropic reinforcement in the top the work equation can be applied as below for unit displacement under P

$$P \times 1 = m \times 2x \times \frac{1}{x}$$

$$P = 2m \qquad \ldots\ldots 3(\text{xiv})$$

Using equation 3(xii) another mode is for a quarter circle fan to develop (Figure 3.12) thus

$$P = (m + 0)\frac{\pi}{2}$$

$$P = 1.57m \qquad \ldots\ldots 3(\text{xv})$$

It is interesting to note that if positive reinforcement is provided then equation 3(xii) can be written as

$$\text{internal work} = m(1 + \mu)\frac{\pi}{2}$$

and to give $P = 2m$ then

$$2m = m(1 + \mu)\frac{\pi}{2}$$

$$\frac{4}{\pi} = 1 + \mu$$

$$\mu = 0.273$$

This case has been examined by Wood[17] who observes that the quarter circle fan solution with top and bottom reinforcement is suspect as it implies that positive radial moments are developed at the free edge of the slab. He suggests leaving a rigid portion next to the slab edges leaving a smaller interior fan (Figure 3.13). This mode is then examined with various values of μ from 0 to 1. With $\mu = 1$ no fan will develop.

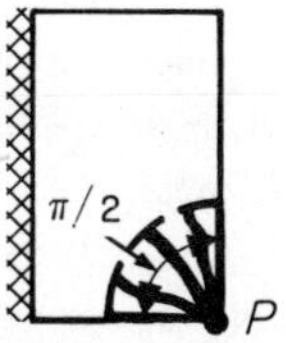

Fig. 3.12 90° fan

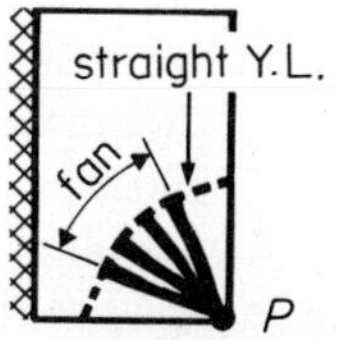

Fig. 3.13 Part fan with rigid portion next to slab edges

Case (*b*), Figure 3.14. As with case (a) equation 3(xii) can be applied to give a semicircular fan, and thus for isotropic top and bottom reinforcement and $\mu = 1$

$$P = m(1 + 1)\pi$$

$$m = \frac{P}{2\pi} \qquad \text{..... 3(xvi)}$$

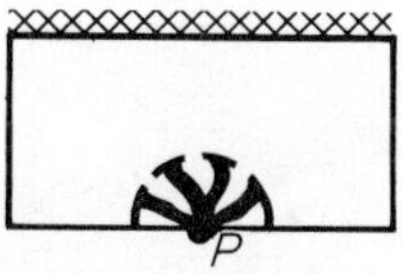

Fig. 3.14 180° fan

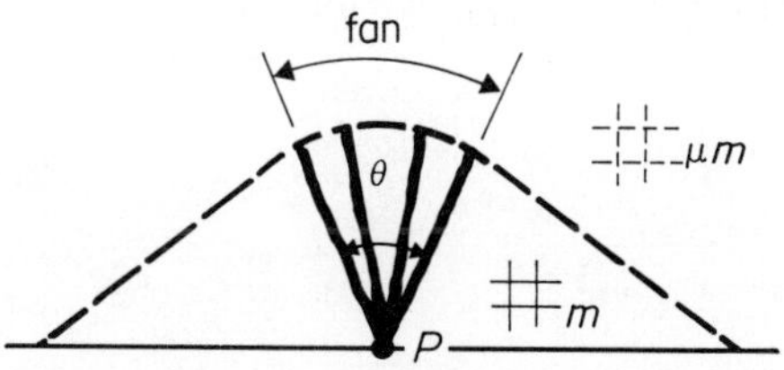

Fig. 3.15 with $\mu = 1$ $\theta = \frac{\pi}{2}$

Again this implies that positive radial moments are applied at the free edge and thus a mode as indicated in Figure 3.15 should be investigated. For $\mu = 1$ Wood[17] shows that $\theta = \pi/2$ and

$$m = \frac{P}{5.14} \qquad \text{..... 3(xvii)}$$

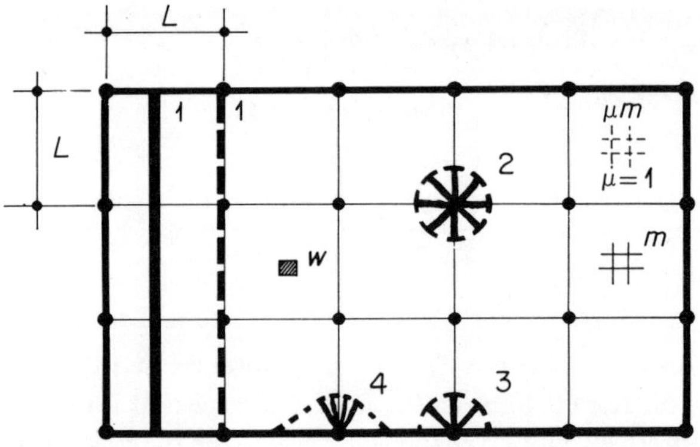

Fig. 3.16 Flat slab on column supports, some possible failure modes with isotropic reinforcement

Example 5, Figure 3.16
A flat slab or plate floor rests on point supports with a grid of $L \times L$ and is subjected to uniformly distributed loading. This type of construction lends itself to analysis by means of the yield line theory but the analysis becomes complicated if varying reinforcement is used. The use of a constant mesh in the top and bottom of the slab simplifies the steelwork fixing and minimizes the possibility of errors in the spacing of the reinforcing bars. The analysis of the floor may be carried out making use of the results obtained previously. Possible yield line patterns are shown in Figure 3.16 and a numerical solution to this problem is given in Chapter 5.

Failure mode 1: $m = \dfrac{wL^2}{11.65}$ end bay

$$m = \frac{wL^2}{16} \quad \text{internal bay}$$

Failure mode 2 (full circular fan):

$$m = \frac{P}{4\pi}$$

where $P \sim wL^2$

Failure mode 3:

$$m = \frac{P}{2\pi}$$

where $P \sim 0.5\, wL^2$

Failure mode 4:

$$m = \frac{P}{5.14}$$

where $P \sim 0.5\, wL^2$

The previous examples have related to slabs with isotropic reinforcement but in a number of situations such as a rectangular slab an orthotropic arrangement gives greater economy.

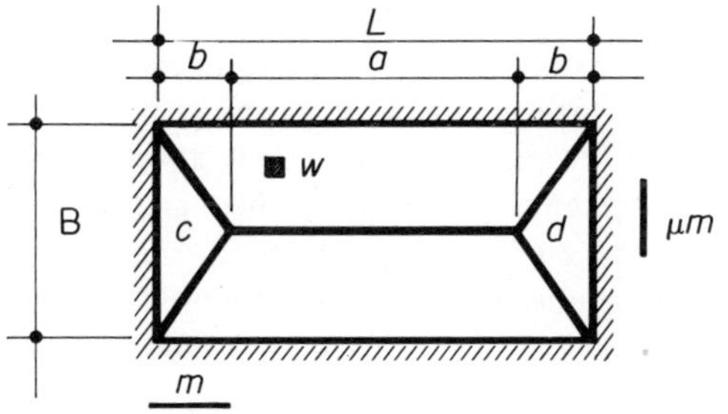

Fig. 3.17 Rectangular slab with orthotropic reinforcement, unit displacement along c d

Example 6, Figure 3.17
A rectangular slab is simply supported at the edges and subjected to uniformly distributed loading. The components of work approach may be used as follows for the postulated yield pattern with unit displacement at centre of slab

$$\text{internal work} = m \times L \times \frac{2}{B} \times 2 + \mu m \times B \times \frac{1}{b} \times 2$$

$$= m\left[\frac{4L}{B} + \mu\frac{2B}{b}\right] \qquad \ldots\ldots 3(\text{xviii})$$

$$\text{external work} = w \times a \times \frac{B}{2} \times \frac{1}{2} \times 2 + \frac{wBb}{3} + \frac{wBb}{3}$$

$$= wB\left[\frac{a}{2} + \frac{2b}{3}\right] \qquad \ldots\ldots 3(\text{xix})$$

Equations 3(xviii) and 3(xix) may be equated and the resulting expression differentiated to find a minimum value of w.

To summarize, the yield line theory is a powerful and elegant method of analysing slabs but from a design office viewpoint there are two 'red lights'.

(a) For other than the simplest configurations and loadings there will be some doubt regarding the use of a failure mode which is not necessarily the critical one – an upper bound solution.
(b) It is not always economic to use isotropic or orthotropic reinforcement with a constant bar spacing. If reinforcement is curtailed then other possible failure modes should be investigated[16] thus further complicating the analysis.

It would be reasonable to state that yield line theory has not been widely used in English design offices but with the recent increase in interest in membrane effects in slabs, giving an enhancement of load carrying capacity, the theory can be applied with greater confidence.

(b) MEMBRANE AND CATENARY ACTION

The partial collapse of a precast concrete structure at Ronan Point[21] has precipitated interest in the post yield response of reinforced concrete slabs to load. Load tests on a three-storey concrete structure in 1955[22] demonstrated that the capacity of lightly reinforced concrete slabs which are contained by stiff beams or additional slabs is in excess of that predicted by the yield line theory. This is known as compressive membrane action and is demonstrated qualitatively in Figure 3.18. Tests carried out on slab strips restrained against longitudinal movement[23] have shown that the maximum load varies from 3 to 17 times that predicted by the yield line theory (bending only), depending on the concrete strength and percentage of reinforcement.

Tensile membrane action is associated with large deformations which could be permitted under accidental impact and explosions and can be developed without heavy edge restraint. Under large deflections, the change in geometry is such that the effective lever arm of the internal forces is amplified – the slab acting as a concave shell with a compressive zone round the perimeter and a tensile zone in the centre with the cracks penetrating to the top surface of the slab[24] (Figure 3.19). If it can be assumed that the total load is carried by catenary action[25] which will exist under large deformation, then a simplified design approach can be adopted. Consider the removal of an internal support to a continuous slab system ABCDE (Figure 3.20). If the dip of the catenary is limited to 10 per cent of the double span $2L$, then the maximum load w is given by

$$\frac{w \times (2L)^2}{8} = A_{st} \times f_y \times \frac{L}{5}$$

$$w = \frac{A_{st} \times f_y}{L} \times \frac{2}{5}$$

$$= 0.4 \frac{A_{st} f_y}{L} \qquad \ldots\ldots 3(xx)$$

This implies continuous bottom reinforcement, and for further information reference should be made to the Building Research Establishment seminar on the stability of precast concrete structures.[25]

(c) THE HILLERBORG STRIP METHOD

An alternative approach to the use of Johansen's yield line theory is the Hillerborg strip method,[26] which is essentially a design method in which the load is carried by a series of strips assumed to act like beams which are reinforced along their full length. This means that the slab may be divided into a

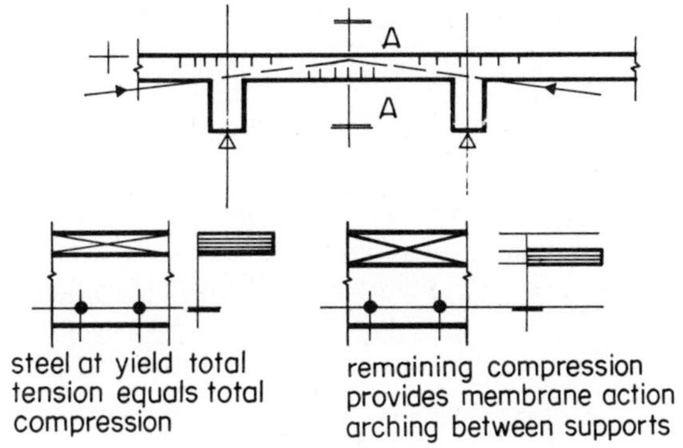

Fig. 3.18 Lightly reinforced slab restrained at edges

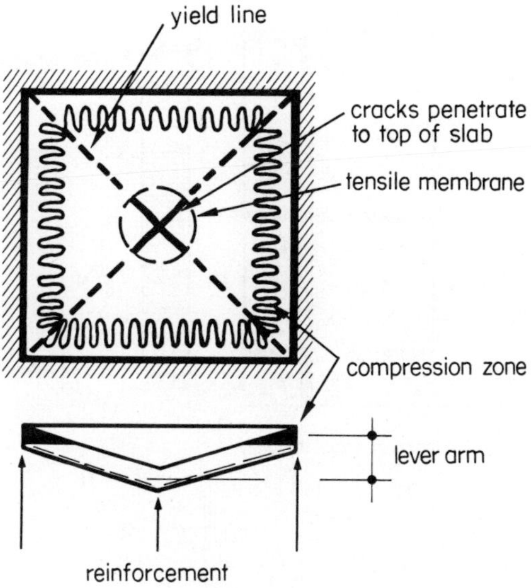

Fig. 3.19

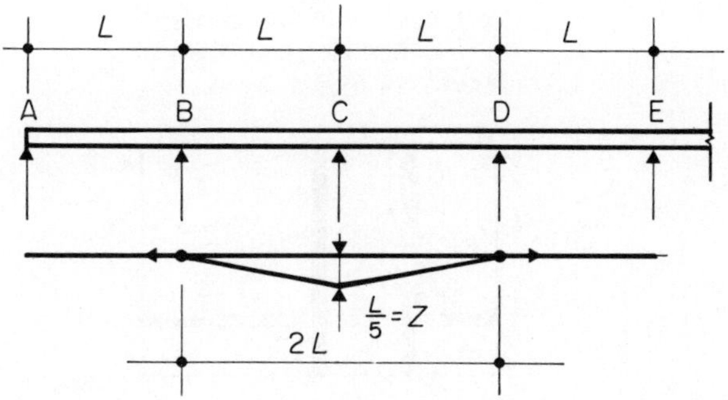

Fig. 3.20

series of X and Y strips thus reducing the equilibrium equation of a slab element to the form

$$\frac{\partial^2 M_x}{\partial x^2} + \frac{\partial^2 M_y}{\partial y^2} = -(w_x + w_y) = -w$$

It is assumed that plasticity allows the deflections of perpendicular strips to become compatible. Division of the load into X and Y strips is achieved by introducing load dispersion lines (Figure 3.21). This implies that triangular or

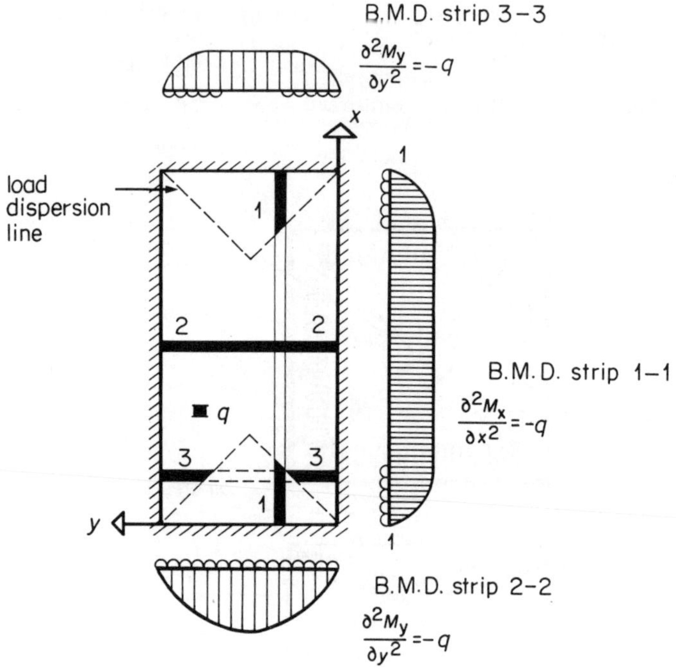

Fig. 3.21

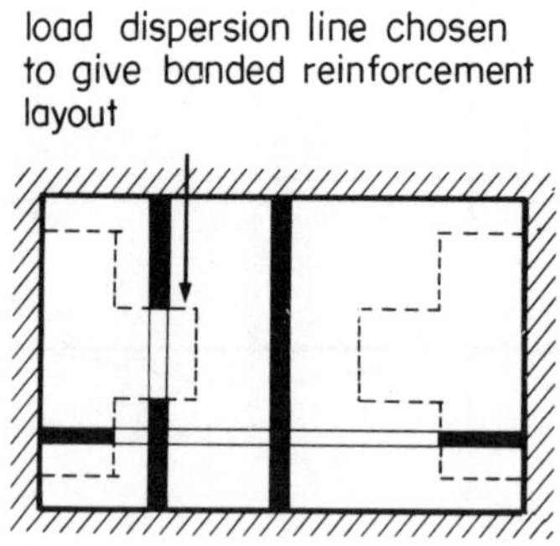

Fig. 3.22

trapezoidal loaded areas are formed which means that continuously varying reinforcement is required, and this is not practical. Wood and Armer[17] suggest the adoption of discontinuity lines as shown in Figure 3.22, and thus the reinforcement may be placed in bands of convenient width.

Continuity over supports is dealt with by introducing zero moment lines (Figure 3.23). The position of the zero moment lines should be chosen so that the ratio of support to span moment does not depart too far from that obtained

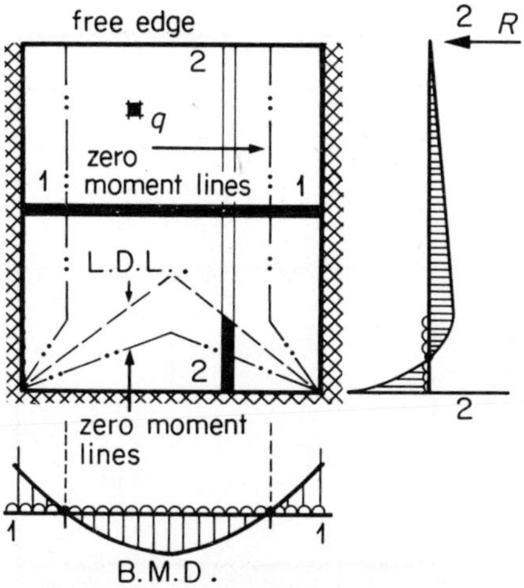

Fig. 3.23 Reaction R required to maintain equilibrium of strip 2–2: additional reinforcement placed in edge to carry reaction R

by an elastic analysis. A large departure may result in excessive steel strain and thus crack width at service load. One problem with the use of the strip method is obtaining a suitable dispersion of concentrated loads to the slab boundaries. Wood and Armer suggest that point loads can be dealt with by the use of strong bands to distribute them to the supports (Figure 3.24). However, the choice of

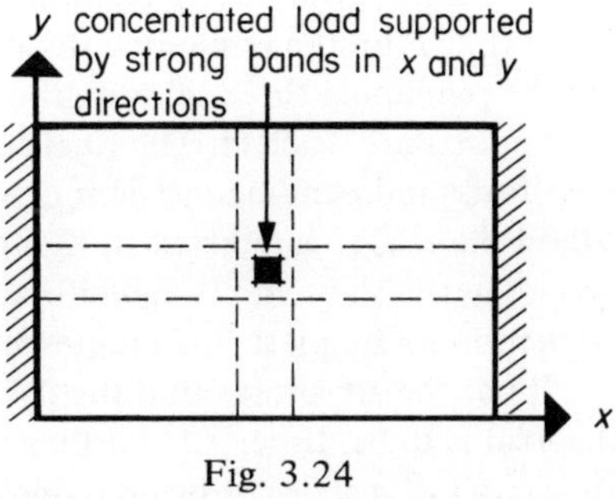

Fig. 3.24

the width of band is difficult to make and if the slab is supported on two sides only some form of edge stiffening will be required to support the reactions at the end of the Y band.

An ingenious generalization of the strip method has been presented by K. O. Kemp[28] in which the designer chooses a distribution pattern for the vertical shear forces rather than the applied loading. At the element where two orthogonal strips intersect, the intersection forces between them must be equal and opposite. Consider a loaded central square of a symmetrical slab which has been divided up into a square grid (Figure 3.25). The load *P* which is assumed to be

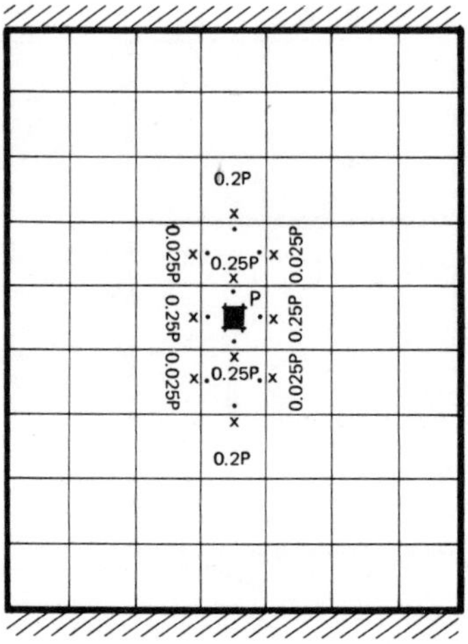

Fig. 3.25 Load *P* on central square balanced by upward forces 0.25 *P* on each side of square. Distribution continued until supports are reached

placed symmetrically on the central square is balanced by upward shear forces of 0.25 *P* on each side. On the adjacent squares there will be a downward shear of 0.25 *P* on one side which, for equilibrium to be maintained, must be balanced by a total upward shear of 0.25 *P* on the remaining three sides. The choice of the distribution of stress on the remaining three sides is arbitrary. The procedure is then repeated until the supports are reached. The final stage is to determine by statics the resulting shear force and bending moment diagrams. For a detailed explanation of the method the reader is referred to Kemp's paper.[28] The method can also be applied to asymmetric slabs and irregular boundary conditions.

Hillerborg's strip method in its simplest form represents a powerful alternative to the yield line theory. It has the advantage that the designer has a free choice of the way in which the load is to be dispersed and thus the design reduces to the consideration of the statics of a series of beam strips, and the distribution of shear along the slab edges is completely defined. It is possible to design a slab with variable reinforcement in contrast to the yield line theory which involves analysing a slab with given uniform reinforcement.

(d) MOMENT REDISTRIBUTION IN BEAMS

The arguments developed previously for the yield line analysis of slabs could equally be applied to beams if the moment/curvature relationship at any section can be idealized to the bi-linear form. This approach will be considered in Section (e).

An alternative procedure is to carry out an elastic analysis of the beam system and then modify the reactant diagram to give a different distribution of moments. This redistribution of moment is in effect an admission of inelastic response to loading at ultimate conditions. As indicated in Chapter 2, CP 114 : 1957 allowed 15 per cent redistribution of moments obtained from an elastic analysis, and for section analysis using the load factor method the neutral axis factor x was limited to $0.5d$. In CP 110 : 1972 the maximum redistribution of moment may be increased to 30 per cent, but the neutral axis depth at this value is limited to $0.3d$. The code states that where, as a result of redistribution, the ultimate resistance moment at a section is reduced, the neutral axis depth x of the section resisting the reduced moment should not be greater than

$$x = (0.6 - \beta_{red})d \qquad \ldots\ldots 3(xxi)$$

where d is the effective depth.

β_{red} is the ratio of the reduction in resistance moment to the numerically largest moment given anywhere by the elastic maximum moments diagram for that particular member, covering all appropriate combinations of ultimate loads. Tabulated values of x for β_{red} up to 0.3 are given:

x	β_{red}
$0.6d$	0
$0.5d$	0.1
$0.4d$	0.2
$0.3d$	0.3

Returning to the two span beam considered in Chapter 2, the maximum moment at support B for an elastic analysis assuming constant flexural rigidity is $\gamma_f wL^2/8$, see Figure 3.26(c). This can be split into two diagrams, the simply supported beam diagram and the reactant diagram, Figure 3.26(b) and (c). If it is proposed to redistribute say 30 per cent of the moment at B, this will reduce to

$$0.7 \times \gamma_f \frac{wL^2}{8} = \frac{\gamma_f wL}{11.42} \quad \text{(Figure 3.26(d))}$$

the reaction at A,

$$RA = \frac{\gamma_f wL}{2} - \frac{\gamma_f wL}{11.42}$$

$$= 0.4125\ \gamma_f wL$$

Thus maximum span moment occurs at $0.4125L$ from the support A and is given by

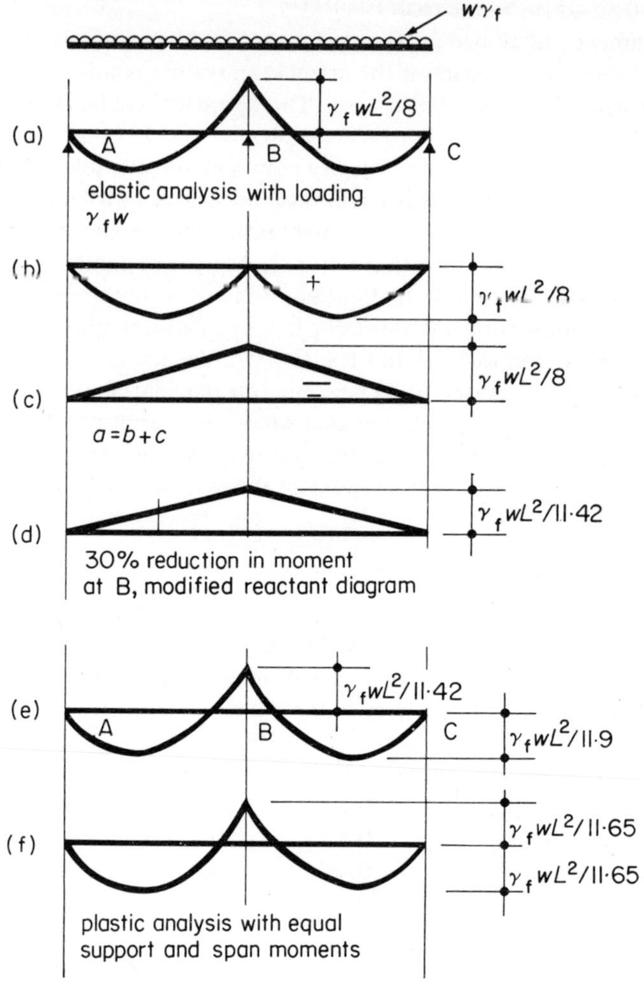

Fig. 3.26

$$M_{\text{span}} = \frac{0.4125^2}{2}\,\gamma_{\text{f}} wL^2$$

$$= 0.085\,\gamma_{\text{f}} wL^2 = \frac{\gamma_{\text{f}} wL^2}{11.9}$$

Thus the final moment distribution is as in Figure 3.26(e). This distribution is very close to that obtained by using a plastic analysis with the ultimate resistance moments in the support and span equal giving

$$M = \gamma_{\text{f}}\,\frac{wL^2}{11.65}, \quad \text{see Figure 3.26(f)}$$

The value of the neutral axis depth will not be known until the section analysis for flexure has been carried out, and thus the percentage distribution assumed initially may require modification depending on the value obtained for x. It is implied in CP 110 that if a plastic analysis is used then the neutral axis depth should be limited to $0.3d$. For slabs this depth is rarely exceeded but for beams with a minimum depth requirement a restriction in neutral axis depth to $0.3d$ can be a disadvantage. Thus it would seem worth while examining the possibility of carrying out a plastic analysis and allowing an increase in the neutral axis depth. This will necessitate making a check on the plasticity of the sections by calculating permissible and actual hinge rotations. This somewhat tedious procedure will be outlined in the next section.

(e) PLASTIC ANALYSIS OF BEAMS

The plastic analysis of reinforced concrete beams was pioneered in this country by A. L. L. Baker, and in 1956 his book on ultimate load theory was published[29] and gave recommendations for the design of concrete beams which incorporated the concept of a plastic hinge and expressions for calculating permissible and actual hinge rotations. The Institution of Civil Engineers report on ultimate load design of concrete structures[12] published in February 1962 also gives expressions for calculating permissible hinge rotations. These expressions have subsequently been modified[30] and their application will be considered later. There is some controversy with regard to the need to calculate hinge rotations, and tests carried out by the Cement and Concrete Association[31] on under-reinforced, balanced and over-reinforced sections indicate that this can be eliminated by the addition of a small quantity of secondary reinforcement. A summary of these results is given below.

(a) Under-reinforced concrete beams probably have more than adequate plasticity at failure and should not require any secondary reinforcement at hinges.

(b) The balanced reinforced concrete beams failed in a rather more brittle manner, but very much improved moment/rotation characteristics can be obtained by the use of secondary reinforcement. This took the form of stirrups or helices. The helices appeared to be more economical than stirrups in terms of weight of steel. Baker has also shown that concrete that is well bound[29] has a limiting strain of several times the unbound value of 0.003 to 0.004.

(c) Over-reinforced concrete beams failed in a very brittle manner and to produce an idealized moment curvature relationship a combination of helices and closely spaced stirrups appeared to be necessary.

(d) In beams in which the main reinforcement was a high percentage of high tensile steel, the effect of helical binding was to increase the maximum moment of resistance of the plastic hinge to a value significantly higher than that obtainable without binding.

The influence of shear on the rotation capacity of reinforced concrete beams has also been investigated[32] and it was found that this influence was marginal for beams designed in accordance with the draft unified code.

Returning to Baker's approach it is assumed that, in a continuous beam system, plasticity will develop at points of maximum stress for elastic conditions which will generally be at the supports. The steel will yield at the supports over a short length of the beam resulting in a so-called plastic hinge. If it is assumed that the section has adequate plasticity, further rotation of the hinge can take place at substantially constant moment (see Figure 2.5). Collapse will occur when a further hinge is formed between the supports. The calculation procedure for checking the rotation capacity of the section is summarized:

(a) Assume a hinge mechanism and a suitable ratio of support to span moment. If this is taken as unity the bending moments for the end span of a continuous system will be as in Figure 3.26.
(b) Calculate the section size and reinforcement required for bending and shear in accordance with the procedure given on pages 71–76 with a maximum neutral axis depth of $x = 0.5d$.
(c) Estimate the permissible and actual hinge rotations using the procedure outlined below. Adequate plasticity is achieved if $\theta_{permissible} > \theta_{actual}$.

The permissible rotation θ_p may be evaluated from the equations given in the ICE report.[12] If S_p is the average strain of the concrete at the compressive edge which occurs under increasing load between the yielding of the steel and the crushing of the concrete, then

$$S_p = e_{cu} - e_{ce}$$

where e_{cu} and e_{ce} are defined on the idealized stress/strain curve shown in Figure 3.27. Thus it is assumed that the steel yields at L_1 (notation as ICE report). If l_p is defined as the length of the member over which the concrete is in effect plastic and θ is the change of slope over the length l_p between steel yield and concrete cracking, Figure 3.28,

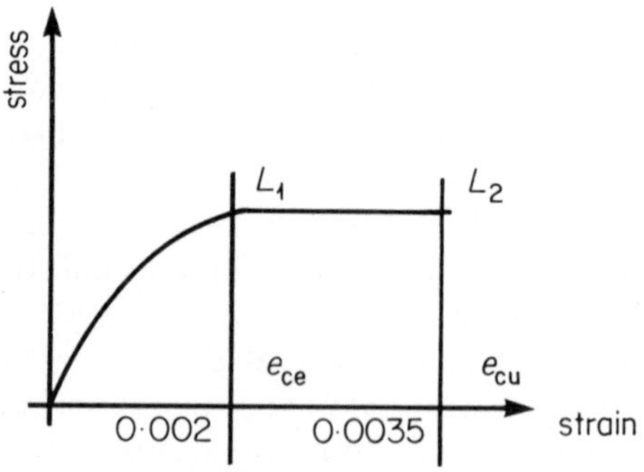

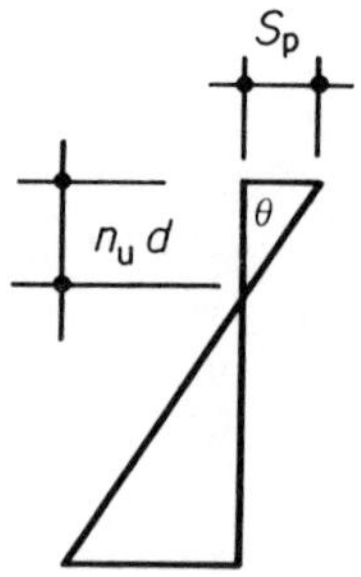

Fig. 3.27 Idealized stress/strain curve for concrete

Fig. 3.28

$$\text{then } \theta = \frac{S_p}{n_u d} = \frac{e_{cu} - e_{ce}}{n_u d} \quad \text{for unit length}$$

$$\text{and over length } l_p, \quad \theta_p = \frac{e_{cu} - e_{ce}}{n_u d} l_p \qquad \ldots\ldots 3(\text{xxii})$$

Equation 3(xxii) represents the rotation on one side of the critical section and l_p is evaluated from the expression[12]

$$l_p = k_1 k_2 k_3 \left(\frac{z}{d}\right)^{\frac{1}{4}} d \qquad \ldots\ldots 3(\text{xxiii})$$

where k_1, k_2, k_3 are parameters relating to the type of steel, presence of axial force and grade of concrete respectively. z is the distance from the critical section to the point of contraflexure and d the effective depth of the section. Recommended values are

$k_1 = 0.7$ mild steel
$k_2 = 1.0$ for no axial load
$k_3 = 0.6$ when cube strength of concrete is 442 N/mm^2
0.9 when cube strength of concrete is 14 N/mm^2
(intermediate values by proportion)

Using some typical figures $k_1 k_3 = 0.5$, $z = 0.15L$ and $L = 15d$ and $n_u d = 0.2d$ then

$$l_p = 0.5 \times (15 \times 0.15)^{\frac{1}{4}} \frac{L}{15}$$

$$= 0.5 \times 1.225 \times \frac{L}{15}$$

Using Figure 3.27

$$e_{cu} - e_{ce} = 0.000\,35 - 0.002$$

$$= 0.0015$$

$$\text{thus from equation 3(xxii)} \quad \theta_p = 0.0015 \times 0.5 \times 1.225 \times \frac{L}{15} \times \frac{15}{0.2L}$$

$$= 0.0015 \times 0.5 \times 1.225 \times 5$$

$$= 0.0046$$

This is the rotation on one side of the critical section and thus

$$\text{permissible } \theta_p = 0.0046 \times 2$$

$$= 0.0092$$

Equation 3(xxii) has subsequently been modified and account is taken of the binding effect of links in the compression zone[30,31]. The inelastic rotation (notation as Reference 30) is given by

$$\theta_p = 0.8\,(e_{c_2} - e_{c_1})k_1 k_3\left(\frac{z}{d}\right) \qquad \ldots\ldots 3(\text{xxiv})$$

$$\text{where } e_{c_1} = 0.0015\left(1 + 1.5p'' + (0.7 - 0.1p'')\frac{1}{n_2}\right) \qquad \ldots\ldots 3(\text{xxv})$$

and p'' is the percentage of binders volume ratio – the total volume of the stirrup excluding any lap length divided by the total volume of concrete; the volume of concrete includes concrete outside the perimeter of the stirrups.

e_{c_1} may be taken as the strain at L_1
and n_2 is the same as n_u in equation 3(xxii).

The presence of links significantly enhances the value θ_p as illustrated in Figure 3.29.

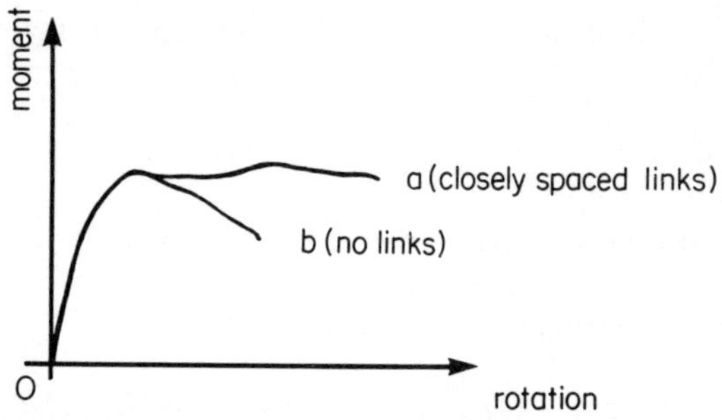

Fig. 3.29 Effect of closely spaced links on moment rotation characteristics of a balanced section

An estimate of the actual hinge rotation can be obtained using the assumption made in the ICE report that the value of EI is constant between the hinges and equal to that at the idealized elastic limit. Use of equation 1(viii) will give a safe minimum value of EI. Thus the actual hinge rotation may be obtained by loading the beam with the M/EI diagram and determining the reactions. From Figure 3.30 with moment coefficients as follows

$$M_B = 0.67\,M_0$$

$$M_c = 0.5\,M_0$$

$$\text{where } M_0 = \frac{wL^2}{8}$$

the hinge rotation at B is given by:

$$\theta_B = \frac{1}{EI}\frac{2}{2} \times M_0 \times \frac{L}{2} \times 2 - \frac{1}{2} \times \frac{2}{3} M_0 \times L \times \frac{2}{3} - \frac{M_0}{2} \times \frac{L}{2} \times \frac{M_0}{6} \times \frac{1}{2} \times \frac{2}{3}$$

$$= \frac{M_0 L}{EI}\;\frac{24}{36} - \frac{8}{36} - \frac{9}{36} - \frac{2}{36}$$

$$= \frac{0.139 M_0 L}{EI} \qquad \ldots\ldots 3(\text{xxvi})$$

It can be seen from the above that a number of dubious assumptions are involved in the estimation of permissible and actual hinge rotations and the procedure is not really suitable for the design office.

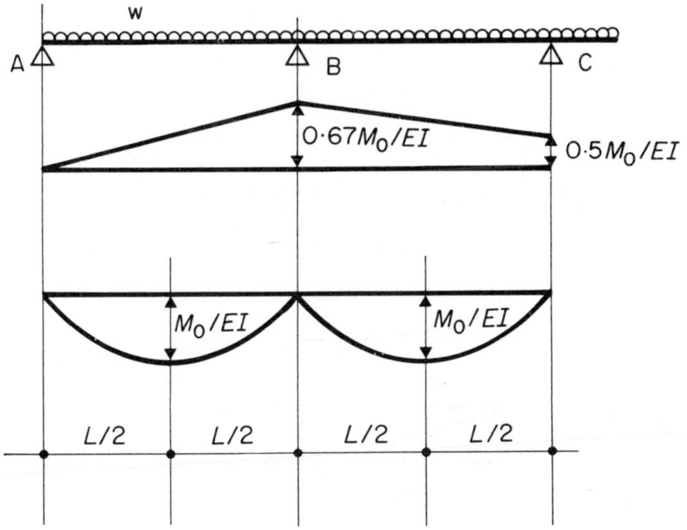

Fig. 3.30 Beam loaded with M/EI diagram to find θ_B. Note $M_O = 0.125\ wL^2$

It is the author's view that a full plastic analysis of continuous beams can safely be carried out allowing utilization of the compression zone in the section analysis of up to $0.5d$. A limited number of tests on two span continuous beams at the University of Surrey have shown that full redistribution of moment is achieved with values of x up to $0.5d$. These beams were designed for shear in accordance with CP 110 but with a maximum link spacing of $0.5d$. Hence the presence of these closely spaced links would enhance the value of θ_p.

(f) ANALYSIS OF COLUMNS

Amplification of moment in columns has been mentioned in Chapter 1, and the treatment of slender columns in CP 110 represents a significant departure from that given in previous British codes of practice. The clauses are based on the CEB recommendations, and the detailed background to their development is given in C & CA research report No. 20.[33] Consider a column with a linear moment/curvature relationship which is assumed not to be influenced by the presence of axial load. From Figure 3.31 for a pin ended column under end moments M_1, the maximum moment is given by

$$M_t = M_1 + N \times a$$

The column will be capable of resisting axial load until the ultimate flexural capacity M_u is reached and thus

$$M_t = M_u = M_1 + N_u \times a_u$$

The variation in curvature is shown in Figure 3.31 and thus a conservative and unconservative approximation may be made. From the equation of bending

$$\frac{1}{r_u} = \frac{M}{EI}$$

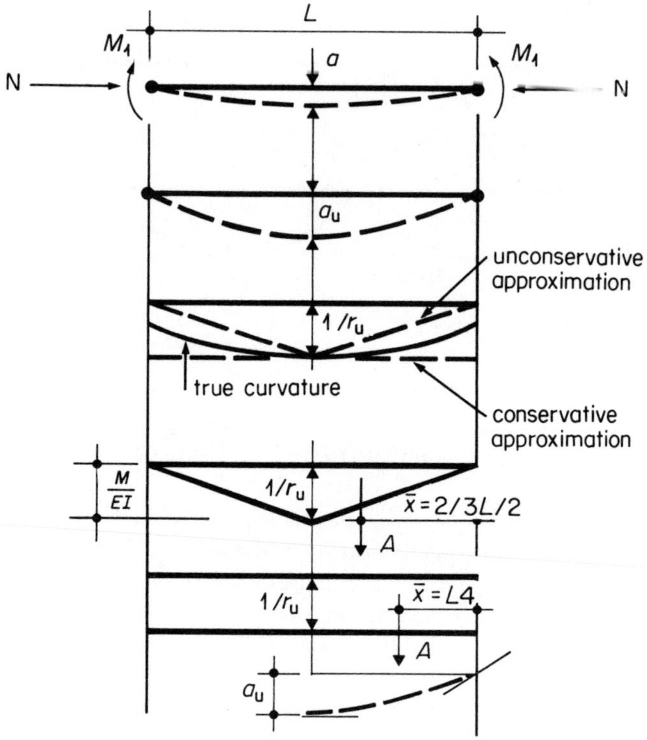

Fig. 3.31

Using the moment area theorem, Figure 3.31, the maximum deflection is given by

$$a_u = \frac{A\bar{x}}{EI}$$

$$\text{unconservative value} = \frac{1}{r_u} \times \frac{L}{2} \times \frac{1}{2} \times \frac{2}{3}\frac{L}{2}$$

$$= \frac{L^2}{12 r_u}$$

$$\text{conservative value} = \frac{1}{r_u} \times \frac{L}{2} \times \frac{L}{4}$$

$$= \frac{L^2}{8r_u}$$

The average value is thus

$$a_u = \frac{L^2}{10r_u}$$

Thus the additional moment may be expressed as

$$M_a = N \times \frac{Le^2}{10} \times \frac{1}{r_u} \qquad \text{..... 3(xxvii)}$$

where Le is the effective column length and $1/r_u$ the curvature at centre of the effective length. Equation 3(xxvii) is identical to that given in the CEB recommendations. To assess the curvature, CEB gives

$$\frac{1}{r_u} = \left[\frac{(0.003 + \frac{f_y{}^*}{E_s} - \frac{L_e}{50\,000h}}{h}\right]k \qquad \text{..... 3(xxviii)}$$

where $f_y{}^*$ = design strength of the reinforcement
h = total depth of section
E_s = modulus of elasticity of reinforcement
k = factor depending on the intensity of axial load.

In CP 110 equation 3(xxviii) is modified, the concrete strain being increased to 0.003 75 to allow for long term effects and $f_y{}^*/E_s$ is taken as 0.002, thus

$$\frac{1}{r_u} = \left[\frac{0.003\,75 + 0.002 - \frac{L_e}{50\,000h}}{h}\right]k$$

This expression simplifies to

$$\frac{1}{r_u} = \frac{1}{175h}\left[1 - 0.0035\frac{L_e}{h}\right]k \qquad \text{..... 3(xxix)}$$

From equation 3(xxviii) and (xxix)

$$M_a = \frac{Nh}{1750}\left(\frac{L_e}{h}\right)^2\left[1 - 0.0035\frac{L_e}{h}\right]k \qquad \text{..... 3(xxxi)}$$

Thus equation 33 in CP 110 may be written as

$$M_t = M_i + \frac{Nh}{1750}\left(\frac{L_e}{h}\right)^2\left[1 - 0.0035\frac{L_e}{h}\right] \qquad \text{..... 3(xxxi)}$$

with k put equal to unity and M_i the initial moment.

The second term in equation 3(xxxi) represents the amplification of moment and may be reduced by a factor $k(\not> 1)$ which represents the influence of axial load. A conservative approach is to put $k = 1$. The factor k is given by

$$k = \frac{N_{uz} - N}{N_{uz} - N_{BAL}} < 1 \qquad \ldots\ldots 3(\text{xxxii})$$

where N = ultimate axial load
N_{uz} = capacity of section under pure axial load
N_{BAL} = axial load corresponding to maximum compressive strain in concrete of 0.0035 and a tensile strain in the outermost layer of steel of 0.002.

As N increases from N_{BAL} upwards the curvature $1/r_u$ and hence the deflection a will decrease. Equation 3(xxxii) allows this to be taken into account but the procedure is complicated by the fact that the steel percentage is unknown at the member analysis stage. Guidance on the evaluation of k is given in the explanatory handbook.[34] In the general case of a column with end moments M_1 and M_2

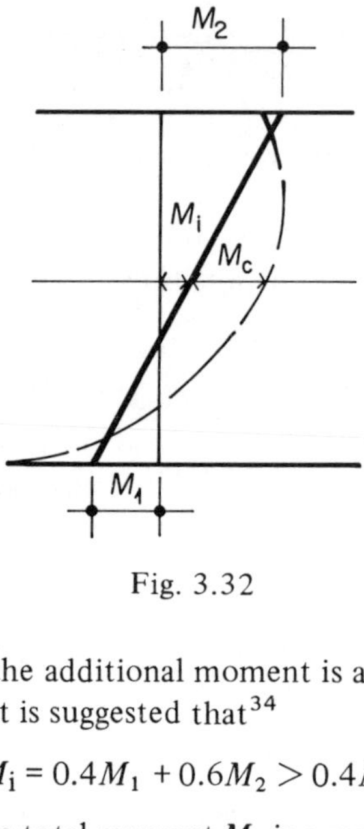

Fig. 3.32

($M_2 > M_1$) (Figure 3.32), the additional moment is a maximum at the central region of the column and it is suggested that[34]

$$M_i = 0.4M_1 + 0.6M_2 > 0.4M_2$$

For a cantilever column the total moment M_t is a maximum at the base. The analysis of sections under combined bending and axial load is included under the general heading section analysis.

Section Analysis

1 FLEXURE

In CP 110 and the CEB recommendations, the concrete stress block is based on a parabolic-rectangular short term stress/strain curve (Figures 3.33 and 3.34). The

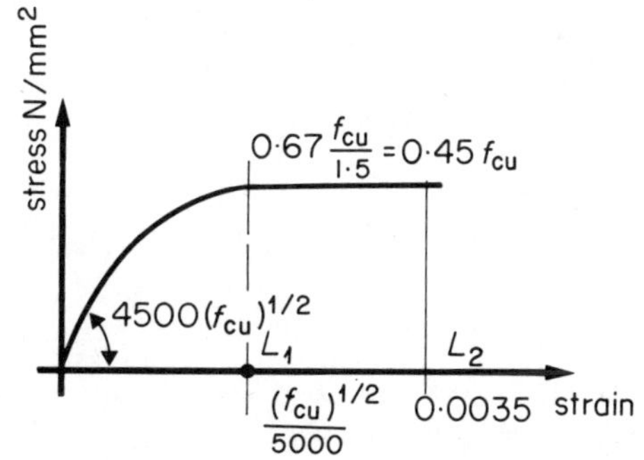

Fig. 3.33 CP110 design stress/strain curve for flexure

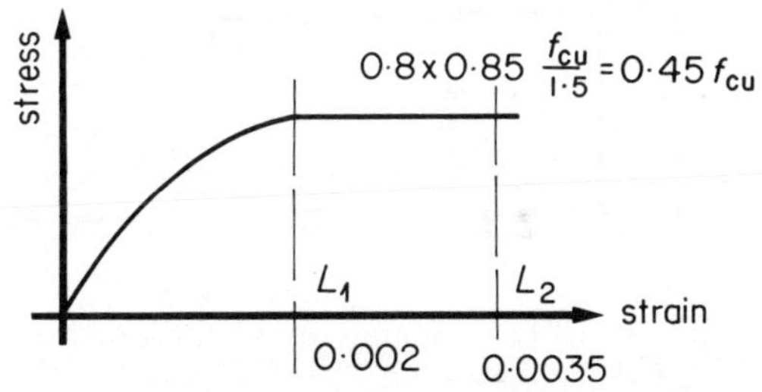

Fig. 3.34 CEB design stress/strain curve for flexure

CP 110 curve has a variable strain at L_1 which is related to the characteristic concrete cube strength f_{cu}. The maximum concrete stress is taken as two-thirds of the cracking strength and with $\gamma_m = 1.5$ for concrete, the design strength approximates to $0.45 f_{cu}$. The CEB curve relates to the characteristic strength of concrete cylinders with a reduction coefficient of 0.85 which allows for:

(a) the difference between short term loads (a few minutes) and loads of a certain duration (a few hours will suffice);
(b) reduction in concrete strength at the extreme fibres due to a more rapid drying out of the surface concrete.

Again with $\gamma_m = 1.5$ and an approximate ratio of 0.8 of cylinder to cube strength, the design strength is

$$0.8 \times 0.85 \times \frac{f_{cu}}{1.5} \sim 0.45 f_{cu}$$

The design curves in Part 2 of CP 110[14] are derived from Figure 3.33, but as an alternative a simplified rectangular stress block may be used with a uniform compressive stress of $0.4 f_{cu}$, Figure 3.35. The value of the neutral axis depth x is limited to $0.5d$ and the resistance moment equations for singly reinforced members are evaluated as follows. From Figure 3.35,

$$M_u = 0.4 f_{cu} b \left(d - \frac{x}{2}\right) x$$

and for $x = 0.5d$ $$M_u = 0.4 f_{cu} b \frac{3}{4} d \frac{x}{2}$$

$$= 0.15 f_{cu} bd^2 \qquad \text{..... 3(xxxiii)}$$

As equation 3(xxxiii) relates to $x = 0.5d$, it should not be used if the amount of redistribution exceeds 10 per cent (see page 59).

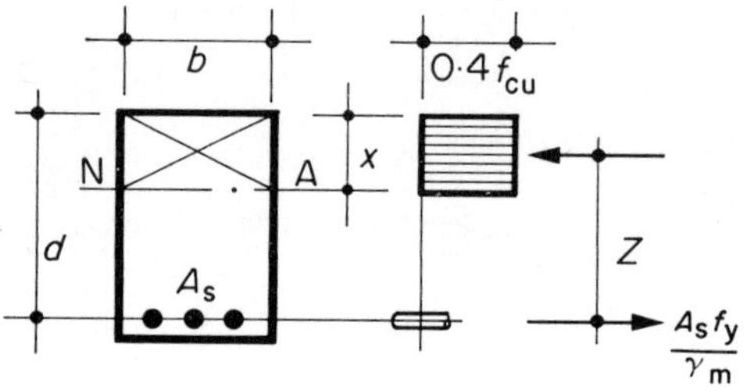

Fig. 3.35 CP110, simplified rectangular stress block

With 30 per cent redistribution, x is limited to $0.3d$ and equation 3(xxxiii) becomes

$$M_u = 0.4 f_{cu} b (d - 0.15d) 0.3d$$

$$\sim 0.1 f_{cu} bd^2 \qquad \text{..... 3(xxxiv)}$$

As indicated previously there are arguments for using equation 3(xxxiii) with redistribution of elastic moments at 30 per cent. Design charts for the simplified rectangular stress block are included in Chapter 5.

The resistance moment can also be expressed in terms of the steel, thus

$$M_u = \frac{f_y}{\gamma_m} A_s z$$

and for $\gamma_m = 1.15 = 0.87 f_y A_s z$ ..... 3(xxxv)

where f_y is the characteristic strength of the reinforcement, the design stress/strain curve for which is shown in Figure 3.36.

If the applied moment at the limit state of collapse exceeds that given by equation 3(xxxiii), the section can be doubly reinforced (Figure 3.37) with

$$M_u = 0.15 f_{cu} bd^2 + 0.72 f_y A'_s (d - d') \qquad \text{..... 3(xxxvi)}$$

and $0.87 f_y A_s = 0.2 f_{cu} bd + 0.72 f_y A'_s$ ..... 3(xxxvii)

In equation 3(xxxvi) the value of $0.72f_y$ is a simplification of

$$\frac{f_y}{\gamma_m + f_y/2000}$$

shown on Figure 3.37. An upper limit of $0.95d$ is used for z.

If the ratio d'/d is large, say greater than 0.2, then the compression steel may be at a lower stress than that implied from equation 3(xxxvi) and the resistance moment should be calculated using a more precise approach making use of Figures 3.33 and 3.36.

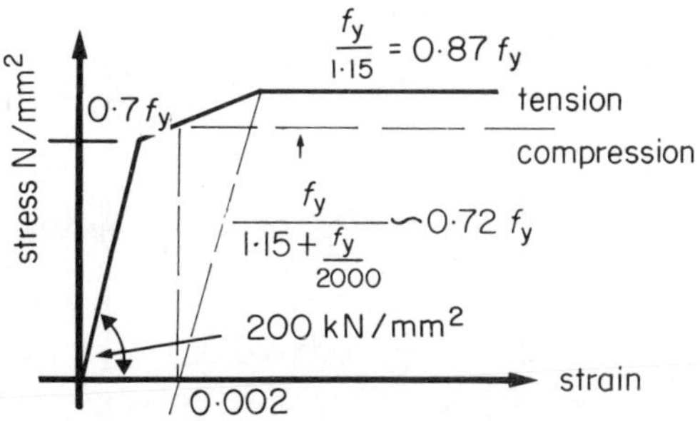

Fig. 3.36 CP110 design stress/strain curve for reinforcement

In the case of T or L beams it is necessary to estimate the effective flange width and CP 110 gives the following rules:

T-beams: the lesser of the width of the web plus one fifth the distance between points of zero moment, or the actual width of the flange.

L-beams: the lesser of the width of the web plus one tenth of the distance between points of zero moment or the actual width of the flange.

From Figure 3.38 the resistance moment may be expressed in the form

$$M_u = 0.4f_{cu}bh_f\left(d - \frac{h_f}{2}\right) \qquad \text{..... 3(xxxviii)}$$

or $$M_u = 0.87f_yA_s\left(d - \frac{h_f}{2}\right) \qquad \text{..... 3(xxxix)}$$

where h_f is the overall depth of the flange.

2 SHEAR

The analysis of reinforced concrete members and sections at the ultimate limit state is considered initially in terms of flexure. This applies to both slabs and beams and the use of an inelastic approach such as the yield line theory or the strip method implies that the designer is satisfied that the composite material has adequate ductility. It is necessary to consider the influence of other force

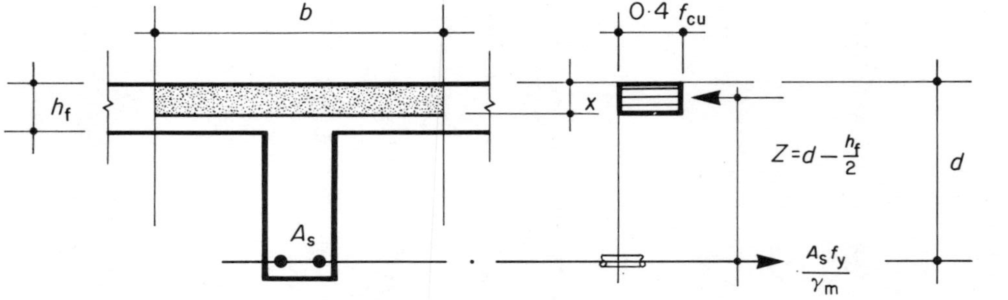

Fig. 3.37

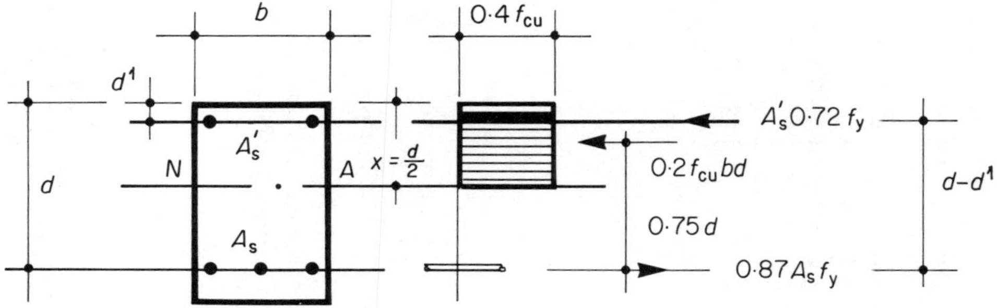

Fig. 3.38

actions on the capacity of the members to deform plastically at ultimate conditions. The influence of shear on the plastic hinge action of reinforced concrete members was investigated by Cranston and Reynolds,[32] and it has been concluded that the rotational capacity is only marginally reduced if shear is designed for in accordance with the CP 110 clauses. A limited number of tests carried out by the author have indicated that adequate rotation capacity can be achieved if shear reinforcement is proportional in accordance with CP 110, but a maximum link spacing of $0.5d$ was used.

In order to take full advantage of the ductility of reinforced concrete members it is necessary to proportion the shear reinforcement such that the load factor for failure in shear is greater than that for flexure. Proposals for achieving this are given in a Building Research Establishment current paper,[35] but the current edition of CP 114 does not give recommendations for ultimate load design with regard to shear. Tests carried out at the BRE[35] have shown that permissible stresses given in CP 114 for beams without shear reinforcement do not give a load factor consistent with that adopted for flexural failure. Another aspect of CP 114 is that shear and flexure are considered separately and their interaction is not taken into account. The analysis of reinforced concrete sections under the effect of bending and shear force actions is a complex problem in which it is difficult to quantify all the factors involved. The forces acting on a

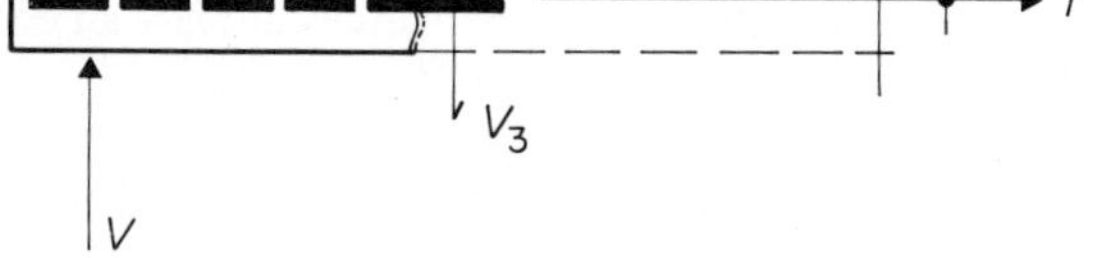

Fig. 3.39 Forces contributing to the transfer of shear across reinforced concrete sections: V_1 concrete compression zone, V_2 aggregate interlock, V_3 dowel action of the longitudinal reinforcement

reinforced concrete section subjected to flexure and shear are shown on Figure 3.39. The factors contributing to the shear resistance are:

(a) the concrete compression zone V_1;
(b) mechanical interlock of the aggregate at the crack V_2;
(c) dowel action of the longitudinal reinforcement V_3.

Attempts[36,37] have been made to assess the relative magnitudes of V_1, V_2 and V_3, and for a detailed treatment of the internal structural action of a reinforced concrete section subjected to shear and the background to the CP 110 clauses reference should be made to the work of P. E. Regan.[38] The influence of V_1, V_2 and V_3 is recognized in CP 110 as account is taken of the contribution of the concrete and the percentage of longitudinal reinforcement.

The shear stress v at any section is calculated from

$$v = \frac{V}{bd} \qquad \text{..... 3(xl)}$$

where V is the shear force appropriate to the ultimate limit state;
b is the breadth of the section (rib width for flanged beams);
d is the effective depth of the section.

If the shear stress v exceeds the appropriate value of v_c given in Table 3.1 (Table 5, CP 110) then reinforcement is provided to resist a total shear of

$$V_R = (v - v_c)bd \qquad \text{..... 3(xli)}$$

Table 3.1 Ultimate shear stress v_c (N/mm^2) in beams

$\frac{100A_s}{bd}$	*Concrete grade (N/mm^2)*			
	20	*25*	*30*	*40+*
0.25	0.35	0.35	0.35	0.35
0.5	0.45	0.50	0.55	0.55
1.0	0.6	0.65	0.70	0.75
2.0	0.8	0.85	0.90	0.95
3.0	0.85	0.90	0.95	1.00

A_s is the area of longitudinal steel which continues at least an effective depth beyond the section considered, thus

$$V = vbd = V_R + v_c bd \qquad \text{..... 3(xlii)}$$

For vertical links at spacing s_v the contribution of the steel over a length d may be written as

$$V_R = A_{sv} 0.87 f_{yv} \frac{d}{s_v} = (v - v_c)bd$$

$$\text{hence} \quad \frac{A_{sv}}{s_v} = \frac{(v - v_c)b}{0.87 f_{yv}} \qquad \text{..... 3(xliii)}$$

where f_{yv} is the characteristic strength of the link reinforcement $\not> 425$ N/mm^2;
A_{sv} is the cross sectional area of the two legs of the link.

If the links are too widely spaced then they may not pick up a potential diagonal crack and in CP 110 the spacing of the legs in the direction of the span and at right angles to it is limited to $0.75d$ maximum.

It should also be noted that CP 110 gives an upper limit of ν (ν_{max}) irrespective of the amount of shear reinforcement provided. This is given in Table 3.2 below (Table 6, CP 110)

Table 3.2 Maximum values of shear stress ν_{max} in beams

	Concrete grade			
	20	*25*	*30*	*40+*
ν_{max} (N/mm^2)	3.35	3.75	4.10	4.75

An upper limit of ν is necessary to ensure that diagonal compressive stresses do not cause crushing of the web concrete. Reference should be made to CP 110 for minimum provision of links, use of bent up bars and the effect of concentrated loads near supports of beams. In the case of slabs which are less than 250 mm deep there is evidence[38] that failure occurs at slightly higher stresses and CP 110 gives an enhancement factor ζ_s for slabs.

Overall slab depth (*mm*)	ζ_s
250 or more	1.0
225	1.05
200	1.10
175	1.15
130 or less	1.20

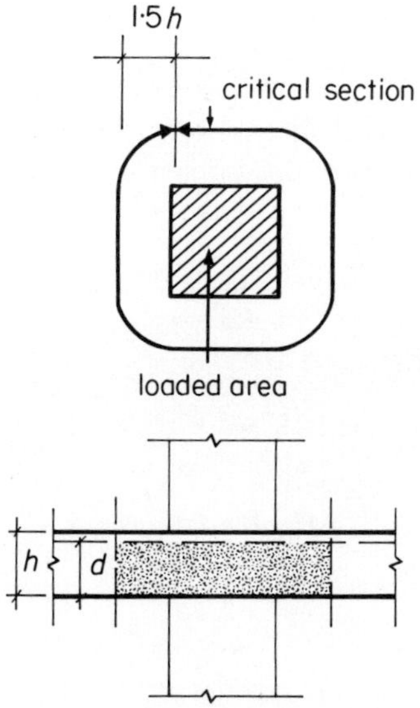

Fig. 3.40

Another important consideration is the effect of concentrated loads on slabs. In the design of flat slab and floors this is often the critical aspect of the design and reference should be made to the example in Chapter 5. In CP 110 the critical section is as shown in Figure 3.40 and the shear stress due to the load is assumed to have a constant value throughout the effective depth and length of the critical perimeter. If the slab is greater than 200 mm thick, shear reinforcement may be employed, but the maximum shear stress due to ultimate loads should not exceed half the value given in Table 3.1 for a beam.

In conclusion the CP 110 clauses differ from CP 114 in the following areas:

(a) shear reinforcement is introduced at a much lower nominal shear stress;
(b) account is taken of the contribution of the concrete and longitudinal steel to the total shear resistance.

With regard to the form of shear reinforcement, tests[39] have indicated that the most efficient types are closely spaced inclusive or vertical links. Bent up bars are the least efficient and the author does not favour their use in general.

3 TORSION

As indicated in Chapter 1 the work of Saint Venant forms a basis for torsion studies, but more recently equations have been developed using the plastic concept of concrete behaviour. It should be noted that plain concrete is neither elastic or plastic (concrete has little plasticity in tension) and some recent information on the mechanism of torsion failure is given in Reference 38.

An equation for the ultimate torsional moment of plain concrete can be developed from the sand heap analogy. If the cross section is cut from a metal sheet or similar, suspended in a horizontal position and covered with sand flowing from an orifice above until it spills over the edges, the natural slope of the sand gives the constant plastic shear stress. The torsional moment T is proportional to twice the volume contained under the sand heap. The volume and slope are dependent on the same constant of proportionality. Considering a square section (Figure 3.41):

$$h = v_{tu} \times \frac{x}{2}$$

$$\text{Twice volume,} \quad 2V = 2 \times \frac{1}{3} \times x^2 \times h$$

$$M_{tu} = v_{tu} \times \frac{x}{2} \times \frac{2}{3} \times x^2$$

$$= \frac{x^3}{3} \times v_{tu}$$

$$\text{Thus} \quad v_{tu} = \frac{3T}{x^3} \qquad \ldots\ldots 3(\text{xliv})$$

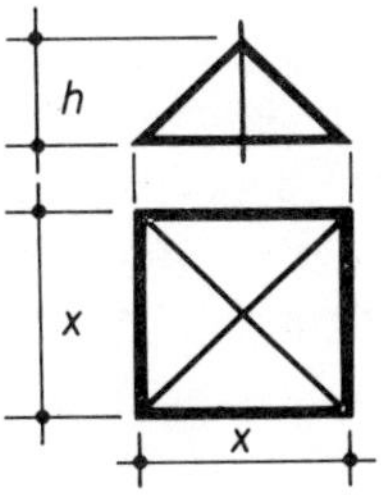

Fig. 3.41

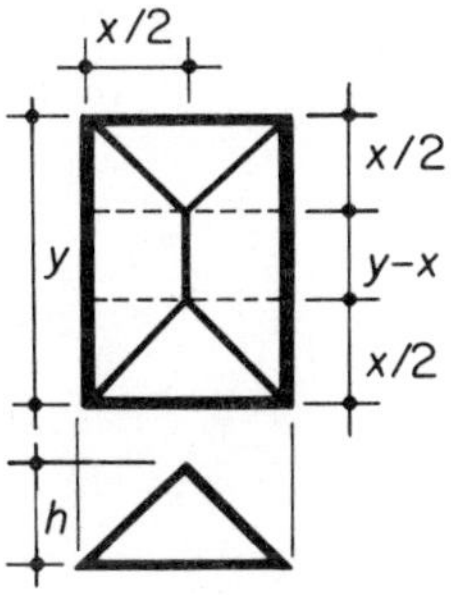

Fig. 3.42

For a rectangular section (Figure 3.42):

$$h = \frac{v_{tu} \times x}{2}$$

$$2V = 2\left[\frac{1}{3}x^2 h + \frac{x}{2}h(y - x)\right]$$

$$T = h\left[\frac{2}{3}x^2 + xy - x^2\right]$$

$$T = V_{tu}\frac{x}{2}\left[xy - \frac{x^2}{3}\right]$$

$$v_{tu} = \frac{2T}{x^2\left[y - \frac{x}{3}\right]} \qquad \text{.....} \; 3(\text{xlv})$$

In practical design, torsion is invariably accompanied by shear and bending. Behaviour in torsion is analogous to behaviour in shear. The resistance of a member to shear or torsion is assumed to be made up of (a) the contribution of the concrete; and (b) the contribution of the web reinforcement.

Limitations are placed on the nominal ultimate shear stress due to torsion and transverse shear. It should be noted that torsional and transverse shear stresses are additive on one side of a beam. The CP 110 recommendations are outlined:

For rectangular sections the torsional shear stress is calculated from

$$v_t = \frac{2T}{h_{min}^2\left[h_{max} - \frac{h_{min}}{3}\right]} \qquad \text{.....} \; 3(\text{xlvi})$$

This equation is identical in form to equation 3(xlv) noting that $h_{min} = x$ and $h_{max} = y$ and T is the torsional moment due to the ultimate loads. Where v_t exceeds $v_{t\,min}$ from Table 3.3 (Table 7, CP 110) reinforcement should be provided and in no case should the sum of the shear stresses resulting from shear and torsion $(v + v_t)$ exceed v_{tu} from Table 3.3.

Table 3.3 Ultimate torsional shear stresses (N/mm^2)

	Concrete grade			
	20	*25*	*30*	*40+*
$\nu_{t\,min}$	0.3	0.33	0.37	0.42
ν_{tu}	3.35	3.75	4.10	4.75

For small sections $y_1 < 550$ mm the value of ν_t should not exceed $\nu_{tu}\, y_1/550$ when y_1 is the larger dimension of the closed link. Torsion reinforcement consisting of rectangular closed links and longitudinal bars should be provided as below.

$$\frac{A_{sv}}{s_v} \geqslant \frac{T}{0.8x_1y_1 0.87f_{yv}} \qquad \ldots\ldots 3(\text{xlvii})$$

$$A_{sl} \geqslant \frac{A_{sv}}{s_v}\frac{(f_{yv})}{(f_{yl})}(x_1 + y) \qquad \ldots\ldots 3(\text{xlviii})$$

where A_{sv} is the area of the legs of the closed links at a section
A_{sl} is the area of longitudinal reinforcement
f_{yv} is the characteristic strength of the links
f_{yl} is the characteristic strength of the longitudinal reinforcement
s_v is the spacing of the links
x_1 is the smaller dimension of the links
y_1 is the larger dimension of the links.

Equations 3(xlvii) and 3(xlviii) are of similar form to 1(xvi) and (xvii) given in Chapter 1 obtained from the thin walled tube approach.

It should be noted that the reinforcement obtained from equations 3(xlvii) and (xlviii) is additional to that required for shear and flexure. The treatment of T, L and I sections is to divide them into component rectangles and summing the contribution of each component.[4]

4 BOND AND ANCHORAGE

The treatment of bond and anchorage is similar to that used for conventional elastic theory thus equations 1(xi) and 1(xii) apply noting that the value of the shear V relates to *ultimate* conditions and that the lever arm z is replaced by the effective depth d (as with shear) thus the local bond stress is given by

$$f_{bs} = \frac{V}{\Sigma u_s d} \qquad \ldots\ldots 3(\text{xlix})$$

for bars which are parallel to the compression face. Limiting values of f_{bs} are given in Table 3.4 (Table 21, CP 110).

Table 3.4 Ultimate local bond stresses (N/mm^2)

Bar type	*Concrete grade*			
	20	*25*	*30*	*40+*
Plain bars	1.7	2.0	2.2	2.7
Deformed bars*	2.1	2.5	2.8	3.4

* can be increased by 20% for ribbed bars.

The anchorage bond stress f_{ba} is obtained by writing equation 1(xii) as

$$f_{ba} = \frac{d}{4L} f_s \qquad \ldots\ldots 3(1)$$

where L is the effective anchorage length of the bar. Where groups of bars are used the effective perimeters of the individual bars are multiplied by a reduction factor which is dependent on the number of bars in the group as below (Table 23, CP 110).

Number of bars in a group	*Reduction factor*
2	0.8
3	0.6
4	0.4

Limiting values of f_{ba} are given in Table 3.5 (Table 22, CP 110).

Table 3.5 Ultimate anchorage bond stress (N/mm^2)

Bar type	*Concrete grade*			
	20	*25*	*30*	*40+*
Plain bar in tension	1.2	1.4	1.5	1.9
Plain bar in compression	1.5	1.9	1.9	2.3
Deformed bar in tension*	1.9	1.9	2.2	2.6
Deformed bar in compression*	2.1	2.4	2.7	3.2

* can be increased by 30% for ribbed bars

5 COLUMNS

Amplification of moments in columns has been considered previously and reference should be made to CP 110 for guidance on the estimation of effective column lengths and the definition of short and slender columns.

Axial loading

The axial load capacity N of a column is based on the stress/strain curves for the concrete and steel, see Figures 3.33 and 3.36. If A_c is the area of concrete A_{sc}

the area of longitudinal reinforcement, than from the design stress/strain curves

$$N = 0.45 f_{cu} A_c + 0.72 f_y A_{sc} \qquad \text{..... 3(li)}$$

A reduction of 10 per cent is applied to equation 3(li) to allow for eccentricity due to construction tolerances and thus

$$N = 0.4 f_{cu} A_c + 0.67 f_y A_{sc} \qquad \text{..... 3(lii)}$$

Equation 3(lii) relates to short braced axially loaded columns and a further reduction factor is applied to cater for moments arising from asymmetrical loading approximately on symmetrical beam arrangements and

$$N = 0.35 f_{cu} A_c + 0.60 f_y A_{sc} \qquad \text{..... 3(liii)}$$

Axial loads plus bending moments

The minimum moment to be considered is that due to the ultimate load acting at an eccentricity of $0.05 h_{min}$, where h_{min} is the minimum depth of cross section. The most practical design office approach is to use the design charts which form parts 2 and 3 of CP 110. These charts apply to sections having a symmetrical arrangement of reinforcement but design formulae are given for rectangular

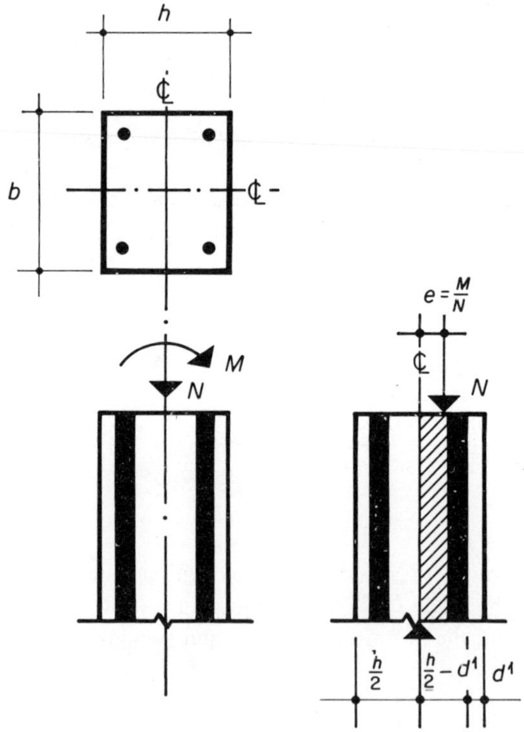

Fig. 3.43

columns and may be used for symmetrical and asymmetrical arrangements of reinforcement. Two cases are considered.

(a) In a column where the ultimate axial load is small and does not exceed the value

$$N = 0.4 f_{cu} b (h - 2e) \qquad \ldots\ldots 3(\text{liv})$$

only nominal reinforcement is required (minimum one per cent) provided eccentricity

$$e = \frac{M}{N} \not< \frac{h}{2} - d'$$

where the notation is as in Figure 3.43 noting that d' is the depth from the surface to the reinforcement in the more highly compressed face.

(b) If the ultimate axial load is greater than the value of N given in equation 3(liv) (equation 27, CP 110) then the strength of the section may be assessed using the equilibrium equations for longitudinal forces and bending moments (equations 28 and 29, CP 110). This is an arithmetically tedious procedure as it is a trial and error approach.

An alternative is to treat the column as a beam by moving the line of action of the axial load to the centroid of the tension steel and designing for the increased moment of

$$N\left(\frac{h}{2} - d_2\right), \quad \text{see Figure 3.44}$$

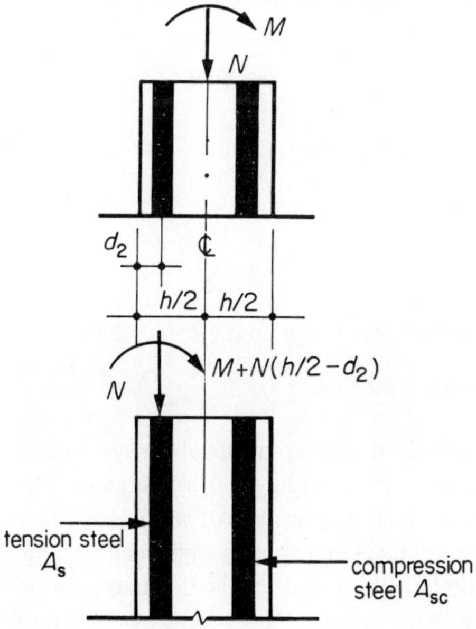

Fig. 3.44 Line of action of N moved to centre of tension steel and moment increased to $M + N\,(h/2 - d_2)$

Thus the design moment is given by

$$M_a = M + N\left(\frac{h}{2} - d_2\right) \qquad \ldots\ldots 3(\text{lv})$$

Equation 3(lv) (equation 30 in CP 110) is applied when the resultant eccentricity e is not less than

$$\left(\frac{h}{2} - d_2\right)$$

The applied moment M_a is equated to the resistance moment of the section which can be expressed in the form obtained for flexural analysis, that is

$$M_a = 0.15 f_{cu} b d^2 + 0.72 f_y (d - d_2) A_{sc} \qquad \ldots\ldots 3(\text{lvi})$$

where A_{sc} represents the amount of compression reinforcement required. As N acts at the centroid of the tension reinforcement the area of tension reinforcement required may be reduced by an amount $N/0.87f_y$. Thus the area of tension reinforcement is given by

$$A_s = \frac{1}{0.87 f_y}\left[0.2 f_{cu} b d + 0.72 f_y A_{sc} - N\right] \qquad \ldots\ldots 3(\text{lvii})$$

This approach has the merit of simplicity as A_{sc} and A_s may be determined directly. However, if the section is designed for reversal of moment, the total steel area obtained becomes uneconomic when compared with that obtained by use of the design charts in part 2 of CP 110.

SUMMARY

To summarize, analysis for the ultimate limit state may be carried out as below.

(1) Members

(a) an elastic analysis with or without redistribution of moment;

(b) an inelastic analysis using the yield line theory or Hillerborg's strip method for slabs, and the plastic hinge concept for beams.

(2) Sections – an inelastic approach

Providing the designer has confidence in the ability of reinforced concrete to respond in a ductile manner to loading at ultimate conditions, an inelastic approach is more logical, and in a number of situations can save time. In particular, the use of the strip method can show a significant saving in design time. However, this can only be achieved if an elastic analysis is not used for the serviceability limit states. A rigorous elastic analysis of complex slab configurations will require the use of a computer and finite element or difference techniques. This can be avoided if the ratio of support to small moment is chosen with reasonable care; CP 110 recommends values between 1.0 and 1.5. An approximation to the elastic distribution of moments at service load can be achieved by taking 70 per cent of those obtained for the ultimate limit state. The format of CP 110 is such that emphasis is placed on detailing rules and limiting span to depth ratios to ensure adequate serviceability rather than direct calculation. However, guidance on methods of compliance with service load criteria by direct calculation is given in the Appendix and this is amplified in the Handbook. The basis for these calculations is the subject matter of Chapter 4.

4

Serviceability Limit States

INTRODUCTION

Concrete is the dominant structural material, and its applications range from long span bridges to the common structures of the urban environment – multi-storey car parks, office blocks, shops, domestic structures, pedestrian subways and lamp posts. No more than a casual inspection of these structures will frequently reveal a number of defects, cracking, staining and spalling, which cast doubt with regard to the durability of concrete. Fortunately, there are many examples of early reinforced concrete work, dating back to the first two decades of the twentieth century and even before, that demonstrate admirably its serviceability and durability. Recently, the author inspected a number of Robert Maillart's[9,40] elegant reinforced concrete bridges. These were constructed over 40 years ago and signs of cracking, spalling and staining are minimal.

Many designers, including the author, have fallen into the trap of proportioning reinforced concrete structures from the criterion of strength, considering dead and imposed loadings only, without regard to the force actions induced by thermal, shrinkage and creep movement. The result of this approach is that the secondary reinforcement is often nominal. A typical example is a long balcony slab designed as a series of cantilever beam strips with nominal transverse reinforcement. From strength considerations this design is adequate, but this is not the case in terms of serviceability, as wide cracks will appear at frequent intervals along the length of the balcony (Figure 4.1). The percentage of transverse reinforcement is inadequate to control the cracking due to thermal/shrinkage effects. Another situation in which excessive cracking can occur is in parapet beams. These are frequently deep in relation to the normal load which they are required to carry and thus may be lightly reinforced. However, an exposed, lightly reinforced beam is vulnerable to thermal effects and thus will require additional reinforcement and movement joints, if the cracks are to be properly controlled.

Some useful guidance on controlling shrinkage and thermal cracking in concrete has been given by B. P. Hughes[41] who emphasizes that a major problem is the contraction which occurs subsequent to the expansion of concrete caused by the heat of hydration of the cement. The evaluation of the spacing of movement joints in concrete structures is dependent on the assessment of creep

Fig. 4 (1) Cracking in a long balcony slab

shrinkage and thermal movements and their interaction. CP 110 does not give values for creep and shrinkage but refers to the CEB recommendations.[13] Some examples of the calculation of the spacing of movement joints, based on the CEB data, are given in a Concrete Society technical paper by J. H. Slack.[42]

CALCULATION OF WIDTHS OF CRACKS

The development of formulae for estimating the surface widths of cracks in the tension zone of sections subjected to flexure has been the subject of a considerable volume of research work,[43] but it must be emphasized that the prediction of an absolute maximum width is not possible. The widths of cracks will vary between wide limits and the possibility of some cracks being wider than those predicted by formulae cannot be avoided.

Prior to the publication of the first draft of CP 110, an investigation,[44] sponsored by the Civil Engineering Research Association, resulted in a proposal (included in the draft) for calculating the maximum surface crack width which could be expressed in the form

$$w = kce \qquad \text{.....}\ 4(i)$$

where w is the crack width

c is the distance from the point of measurement of the crack to the nearest reinforcing bar

e is the surface strain at the level at which the crack width is being calculated

k is a factor to give a crack width with some specified probability of being exceeded. This factor was taken as 3.3 for deformed bars and 4.0 for plain round bars. In the draft this probability factor was reduced to 2.3.

The background to this proposal is given in C & CA Research Report No. 18[44] and its application is outlined:

For assessment of crack widths, the value of γ_f is taken as unity and the bending moment M_e at any section in the structure is obtained from an elastic analysis. The steel percentage is known from the ultimate limit state analysis and thus the neutral axis depth x may be obtained from equation 1(ii) as shown.

$$\text{steel percentage } p = \frac{50}{\alpha e}\,\frac{n^2}{1-n}$$

$$\text{and} \qquad x = nd$$

In order to allow for creep, the appropriate value of the modulus of elasticity of the concrete is half the short term elastic modulus, which may be taken from Table 4.1 (Table 1, CP 110).

Table 4.1 Values of short term modulus of elasticity of concrete

Cube strength of concrete at appropriate age or stage considered (N/mm²)	*Modulus of elasticity of concrete, E_c(kN/mm²)*
20	25
25	26
30	28
40	31
50	34
60	36

Noting that γ_m is taken as 1.3 for assessing crack widths in a structure then

$$\alpha e = \frac{2E_s}{E_c} \qquad \text{..... 4(ii)}$$

where E_s has a constant value of 200 kN/mm^2.

From equation 1(ii) or Table 1.1 the value of n and thus x may be obtained using the value of αe calculated from equation 4(ii). Thus the steel stress f_s may be obtained from the equation

$$f_s = \frac{M_e}{A_s z} \qquad \text{..... 4(iii)}$$

where A_s is the area of tension steel obtained from the ultimate limit state analysis

z is the lever arm $\left(d - \frac{x}{3}\right)$, elastic analysis.

The corresponding strain in the reinforcement is given by

$$e_s = \frac{f_s}{E_s}$$

The draft CP 110 imposed an upper limit on e_s of $0.75f_y/E_s$. Referring to Figure 4.2 the crack width at point 1 is given by

$$w = kc_1 e_1$$

$$= kc_1 e_s \frac{h-x}{d-x}$$

$$= kc_1 \frac{f_s}{E_s}\frac{h-x}{d-x} \qquad \ldots\ldots 4(\text{iv})$$

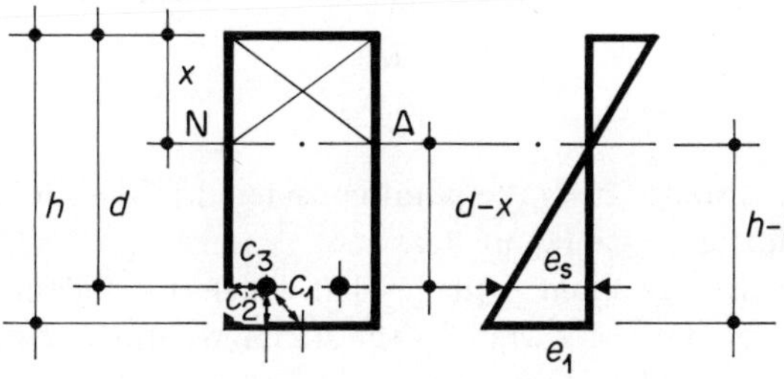

Fig. 4.2

Equation 4(iv) ignores the stiffening effect of the concrete in the tension zone and this will reduce the steel strain. This was allowed for in the draft CP 110 by reducing strain e_1, calculated on the basis of zero tensile strength, by

$$\frac{4bd\,10^{-6}}{A_s}$$

Thus the net strain is

$$e_1 - \frac{4bd\,10^{-6}}{A_s}$$

$$\text{and thus } w = kc_1\left[\frac{f_s}{E_s}\frac{h-x}{d-x} - \frac{4bd\,10^{-6}}{A_s}\right] \qquad \ldots\ldots 4(\text{v})$$

It was found that equation 4(v) worked reasonably well for beams in which c was not greater than about 80 mm, but for slabs the value of c can be much higher and large crack widths will be obtained which do not tally with observations. Thus further tests were carried out on slabs[45] from which the crack width formula given in CP 110 was derived (Appendix A3). The width of flexural cracks depend primarily on three factors.

(a) The proximity to the point considered of reinforcing bars perpendicular to the crack.
(b) The proximity of the neutral axis to the point considered.
(c) The average surface strain at the point considered.

The formula given in CP 110 (equation 61) is expressed as

$$\text{design surface crack width} = \frac{3a_{cr}e_m}{1 + 2\dfrac{a_{cr} - C_{min}}{h - x}} \qquad \ldots\ldots 4(\text{vi})$$

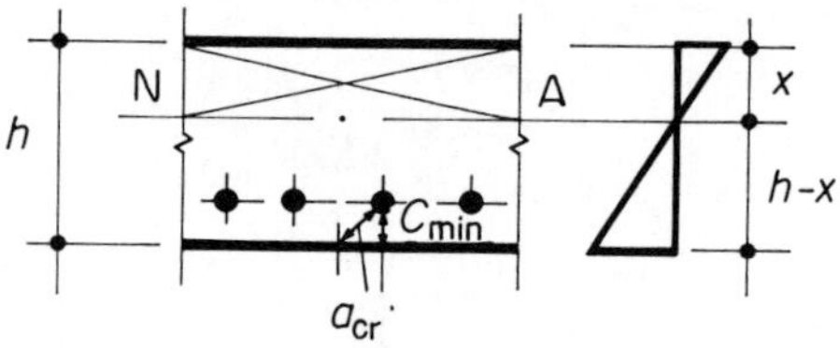

Fig. 4.3

where a_{cr} is the distance from the point considered to the surface of the nearest longitudinal bar, see Figure 4.3.
e_m is the average strain at the level at which the cracking is being considered, calculated allowing for the stiffening effect of the concrete in the tension zone and is obtained from equation 4(vii).
C_{min} is the minimum cover to the tension steel.
h is the overall depth of the member.
x is the depth of the neutral axis found from the analysis to determine e_1 (see below).

$$e_m = e_1 - \frac{1.2b_t h(a^1 - x)}{A_s(h - x)f_y} \times 10^{-3} \qquad \ldots\ldots 4(\text{vii})$$

(equation 62, CP 110)

where e_1 is the strain at the level considered in Figure 4.2, calculated ignoring the stiffening effect of the concrete in the tension zone, that is $\dfrac{f_s}{E_s}\dfrac{h-x}{d-x}$.
b_t is the width of the section at the centroid of the tension steel.
a^1 is the distance from the compression face to the point at which the crack width is being calculated, and
A_s is the area of tension reinforcement.

The application of equations 4(i) to 4(vii) is demonstrated in Chapter 5. It should be noted that in the final version of CP 110 data for the calculation of crack widths has been relegated to the appendix and that the detailing rules given in the main body of the document will, in general, ensure adequate serviceability. However, if the designer adopts an inelastic approach to member and section analysis, an estimate of crack widths may, in some situations, be prudent. In an internal non-aggressive environment an assessment of crack widths should show that a value of 0.3 mm would not in general be exceeded. In an aggressive environment, or where leakage from water retaining structures is to be prevented, this value should be reduced to possibly 0.1 mm. In the case of water retaining structures equation 4(vi) can be used for flexure, and for pure tension a more accurate formula is required, as indicated in the CP 110 Handbook.

Cracking in reinforced concrete structures is inevitable – micro cracks occur even before external loads are applied. Having accepted this the engineer should ensure that cracking is controlled by careful attention to detailing the reinforcement and the provision of movement joints. Time spent on this aspect of the design process will be more beneficial than an elaborate mathematical analysis of an idealized 'elasto-plastic continuum'. It is worth noting that over the past 40 years steel stresses at service load have increased by 300 per cent, indicating the need for much greater attention to crack control – it can be seen from equation 4(iv), that the steel stress is a primary factor governing crack widths.

DEFLECTION

Table 2.1 demonstrates that codes of practice are paying increasing attention to the control of deflection in reinforced concrete members. This has been necessitated by the increase in steel and concrete stresses at service load resulting in shallower sections and reduced flexural rigidity. It has been general practice to avoid calculation of the deflection of reinforced concrete members by using tables of permissible span to depth ratios. The values given in CP 114 : 1957 were as follows:

Beams	*Ratio of span to overall depth*
simply supported beams	20
continuous beams	25
cantilever beams	10
Slabs	
slabs spanning in one direction, simply supported	30
slabs spanning in one direction, continuous	35
slabs spanning in two directions, simply supported	35
slabs spanning in two directions, continuous	40
cantilever slabs	12

These values have been shown by experience to be satisfactory but this experience was largely associated with relatively low tensile steel stresses – 125 N/mm^2 for beams and 185 N/mm^2 for slabs.

In the 1965 revision to CP 114 the use of the above table is limited to members with steel stresses not greater than 140 N/mm^2 and concrete stresses not greater than 10.5 N/mm^2. If one or both of these figures are exceeded (for normal density concrete 2400 kg/m^3), then a reduction factor is applied.

The deflection of reinforced concrete members is influenced by a number of factors which are difficult to quantify in a precise manner, in particular the evaluation of the proportion of the total load which may be considered as permanent, the variation in flexural rigidity along the length of a member which may be cracked or uncracked, of rectangular, T or L section with a variable reinforcement areas (compression and tension), the stiffness of the supports and the influence of creep shrinkage and thermal effects. Some recognition of this is given in CP 110 by introducing tables of *basic* span/effective depth ratios, to which modification factors can be applied. The basic span/effective depth ratios given in Table 4.2 (Table 8, CP 110) are based on limiting the deflection to span/250. This should normally avoid damage to finishes or partitions for spans up to 10 m.

Table 4.2 Basic span/effective depth ratios for beams

Support conditions	*ratio*
cantilever	7
simply supported	20
continuous	26

For spans exceeding 10 m, CP 110 gives the following table (Table 9) to be used when it is necessary further to restrict the deflection to avoid damage to finishes and partitions.

Table 4.3 Special span/effective depth ratios for rectangular beams

Span	*Cantilever*	*Simply supported*	*Continuous*
10	to be justified by calculation	20	26
12		18	23
14		16	21
16		14	18
18		12	16
20		10	13

As indicated previously, the deflection of a reinforced concrete member is influenced by the amount of tension and compression steel, and the code takes this into account by introducing modification factors (Tables 10 and 11, CP 110) to be applied to the basic and special span/effective depth ratios given previously. These factors relate to the percentages of compression and tension steel and the service stress in the tension steel.

The span/effective depth ratios obtained from the above procedure take account of normal creep and shrinkage deformation. An alternative, but generally more laborious procedure, is to assess the deflection by direct calculation.

Calculations Of Deflections

As indicated in Chapter 2 the deflection at any section in a beam system may be expressed as (equation 2(xi))

$$a = K_1 \frac{WL^3}{EI}$$

The difficulty of applying this equation is to obtain a realistic assessment of the value of the flexural rigidity EI. A possible approach is to determine the bending moment M_c at which cracking occurs and then use the gross and cracked section properties for determining the value of the flexural rigidity, as shown in Figure 4.4. By applying unit load at the point at which the deflection is to be calculated, the deflection is determined from the product integration

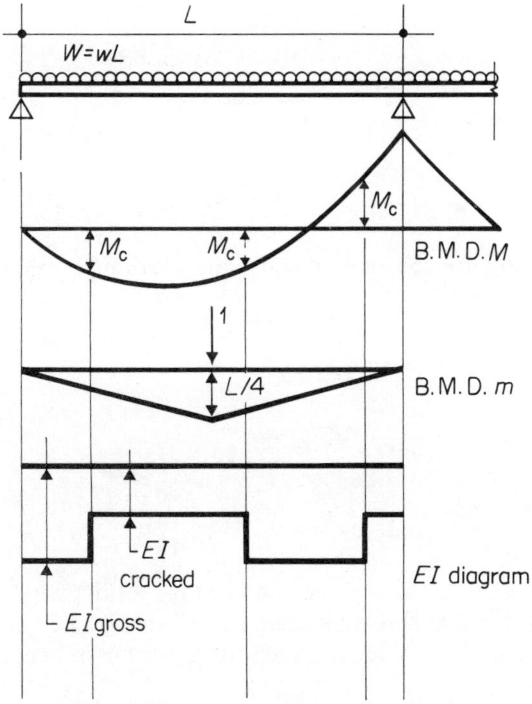

Fig. 4.4

$$a = \int_0^L \frac{M_m \, dx}{EI}$$

However, this approach is tedious and does not take into account the stiffening effect of the concrete in the tension zone, which will increase the stiffness of the section. Equation 2(xii) forms the basis for the method of calculating deflections given in CP 110, which is developed further in the Handbook. Equation 2(xii) gives

$$a = KL^2 \frac{1}{\gamma_b}$$

where $K = \dfrac{K_1}{K_2}$

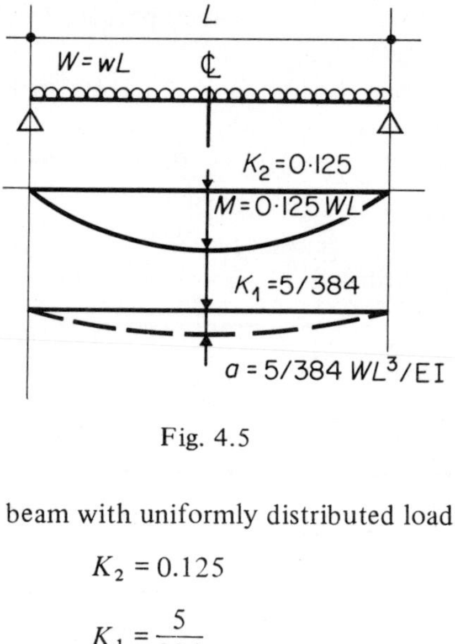

Fig. 4.5

For a simply supported beam with uniformly distributed loading (Figure 4.5),

$$K_2 = 0.125$$

$$K_1 = \frac{5}{384}$$

Thus $$\frac{K_1}{K_2} = \frac{40}{384} = 0.104$$

and $$a = 0.104 L^2 \frac{1}{\gamma_b}$$

In the Handbook,[34] values of K are tabulated for various bending moment diagrams. Thus the calculation reduces to a reasonable assessment of the curvature $1/\gamma_b$ at midspan. As a starting point the equation of bending gives

$$\frac{M}{I} = \frac{E}{\gamma_b} = \frac{f}{y} \qquad \ldots\ldots 4(\text{viii})$$

$$\text{and } \frac{1}{\gamma_b} = \frac{f}{Ey} \qquad \ldots\ldots 4(\text{viii})$$

For a cracked reinforced concrete section, using the notation as in Figure 4.6, $y = d - x$ and

$$\frac{1}{\gamma_b} = \frac{f_s}{E_s(d - x)} \qquad \ldots\ldots 4(\text{ix})$$

where E_s is the modulus of elasticity of the steel.

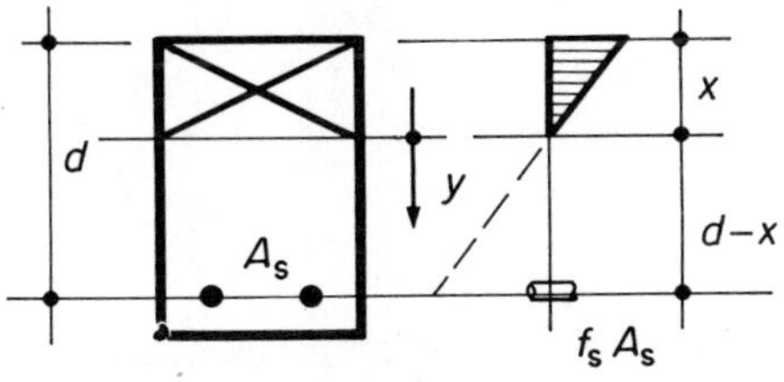

Fig. 4.6

An approximation to the value of x, ignoring the effect of the concrete in the tension zone, is to use equation 1(ii) with $x = nd$, or Table 1.1. From the value of x the steel stress f_s may be determined from equilibrium of moments and thus the value of $1/\gamma_b$ from equation 4(ix).

The equilibrium of moments equation allows for the stiffening effect of the concrete in the tension zone by assuming a short term stress of 1 N/mm^2 (tension)

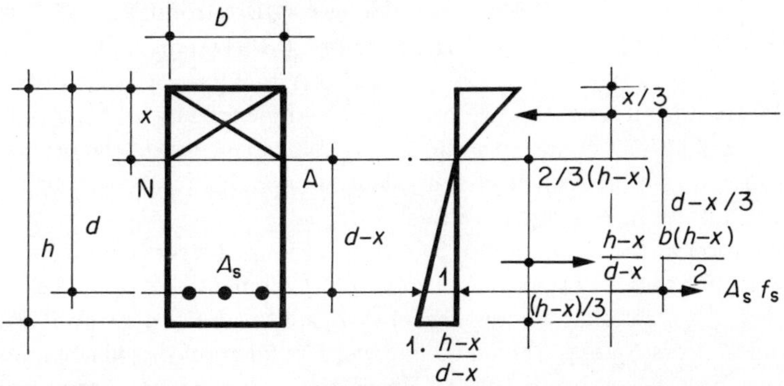

Fig. 4.7

reducing in the long term to 0.55 N/mm^2, see Figure 4.7. Taking moments about the centroid of the compression zone

$$M = A_s f_s\left(d - \frac{x}{3}\right) + 1\,\frac{h-x}{d-x}\,\frac{(h-x)}{2}\left[\frac{2}{3}(h-x) + \frac{2}{3}x\right]$$

$$= \frac{pbd}{100} f_s\left(d - \frac{x}{3}\right) + 1\,\frac{h-x}{d-x}\,\frac{b}{3}(h-x)x$$

$$= \frac{pbd}{100} f_s \left(d - \frac{x}{3}\right) + 1 \frac{bh(h-x)^2}{3(d-x)} \qquad \ldots\ldots 4(x)$$

where $p = \frac{A_s}{bd} 100$

Thus f_s may be determined from equation 4(x) noting that this applied to short term loading with the tension limited to 1 N/mm². If the contribution of the concrete in the tension zone is ignored, then the steel stress is given by

$$f_s = \frac{M}{A_s\left(d - \frac{x}{3}\right)} = \frac{M}{A_s z} \qquad \ldots\ldots 4(xi)$$

The expression

$$1 \frac{bh(d-x)^2}{3(d-x)}$$

represents the contribution of the tensile concrete and can significantly reduce the value of f_s obtained from equation 4(xi), as demonstrated in Chapter 5 (Example 5.02).

The deflection is calculated from the expression

$$a = KL^2 \frac{f_s}{E_s(d-x)} \qquad \ldots\ldots 4(xii)$$

The effect of creep can be taken into account by the introduction of a creep coefficient (see Table A.1 in the Handbook or CEB recommendations) from which a modified value of αe is obtained. This is used in conjunction with the long term tensile stress of 0.55 N/mm² (see Example 5.02).

Increase in curvature due to shrinkage will be dependent on the percentages of tension and compression steel and guidance on evaluation of the shrinkage curvature is given in Appendix A of CP 110.

Calculation of the deflection of one-way slabs may be carried out on a beam strip basis. For two way slabs, tables of bending moment coefficients are available,[6] but it should be noted that they relate to a homogeneous isotropic continuum. A cracked reinforced concrete slab is far from this idealization, but an approach suggested in the CP 110 Handbook is to consider a strip of unit width, chosen so that the moment along it is the maximum moment in the slab, see Figure 4.8. The deflection is then calculated in accordance with the procedure described previously. In the absence of tables of bending moment coefficients 70 per cent of the moments used for the ultimate limit state may be adopted.

To summarize, compliance with serviceability criteria by direct calculation of crack widths and deflections will involve additional design time, and for normal building structures the corresponding increase in design costs are difficult to justify.

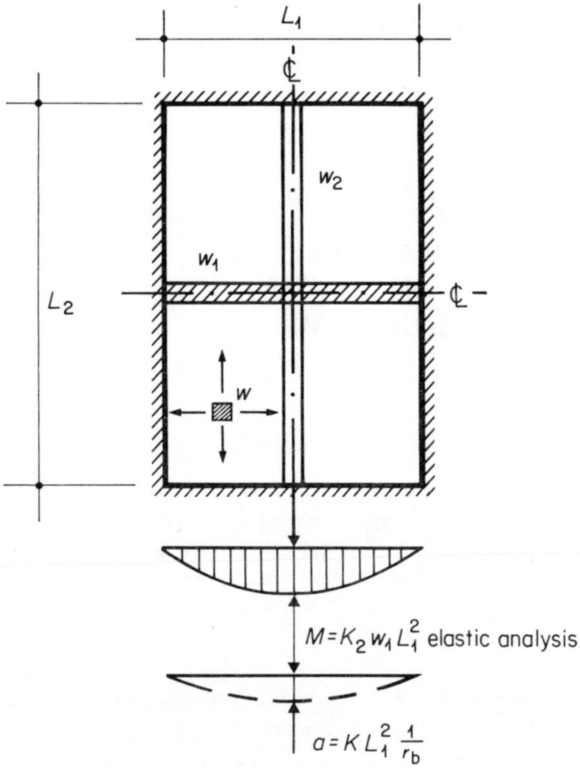

Fig. 4.8 Approximate procedure for calculating deflection of 2-way slabs

Example 5.02 in Chapter 5 illustrates this point, and a number of engineers have expressed serious doubts regarding the validity of CP 110 as a practical document. It has been stated that the use of yield line theory or the Hillerborg strip method is unnecessarily complex for 95 per cent of jobs.[46] The author is inclined to take the opposite view. Strength analysis using these methods will, in many situations, reduce design time, but only if serviceability checks by direct calculation are avoided. This can be achieved by careful attention to the choice of support to span moment ratios and the avoidance of large bar spacings.

5

Typical Calculations

'Nevertheless we must never forget that they (mathematics) only provide us with the means of changing the form of data which we already possess and whatever may be the value of these transformations, we never find, at the end of our calculations more than we put into them at the beginning.'

Eugene Freyssinet

The structural calculation is a confirmation process which is carried out to satisfy the designer that his initial conception of the form and proportions of the structure is adequate in terms of strength and serviceability. The following calculations illustrate in quantitative terms the application of the limit state theory developed in the previous chapters to the analysis of simple structural elements, the beam, slab and column.

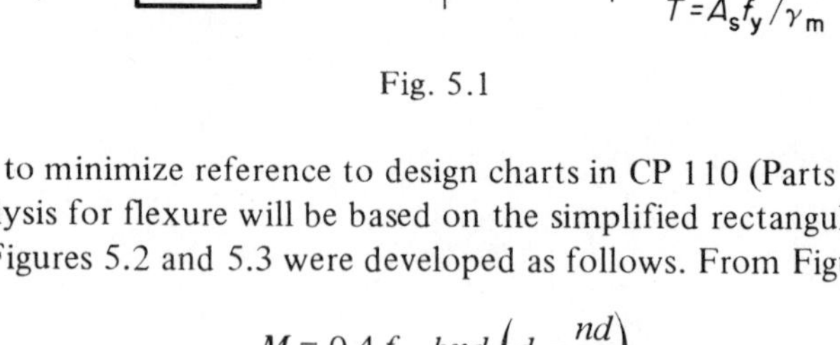

Fig. 5.1

In order to minimize reference to design charts in CP 110 (Parts 2 and 3), section analysis for flexure will be based on the simplified rectangular stress block and Figures 5.2 and 5.3 were developed as follows. From Figure 5.1

$$M = 0.4 f_{cu} bnd \left(d - \frac{nd}{2}\right)$$

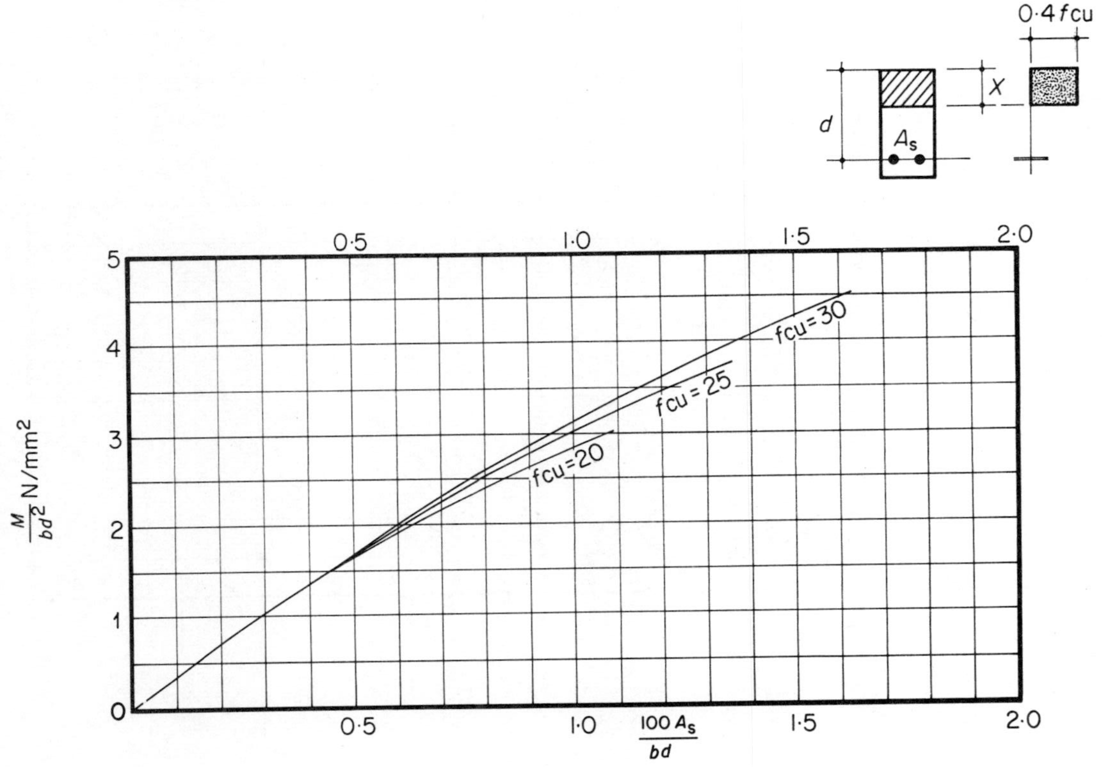

Fig. 5.2 Design chart for singly reinforced beams f_{cu} = 20, 25, 30 N/mm^2, f_y = 425 N/mm^2 $X \not> 0.5d$

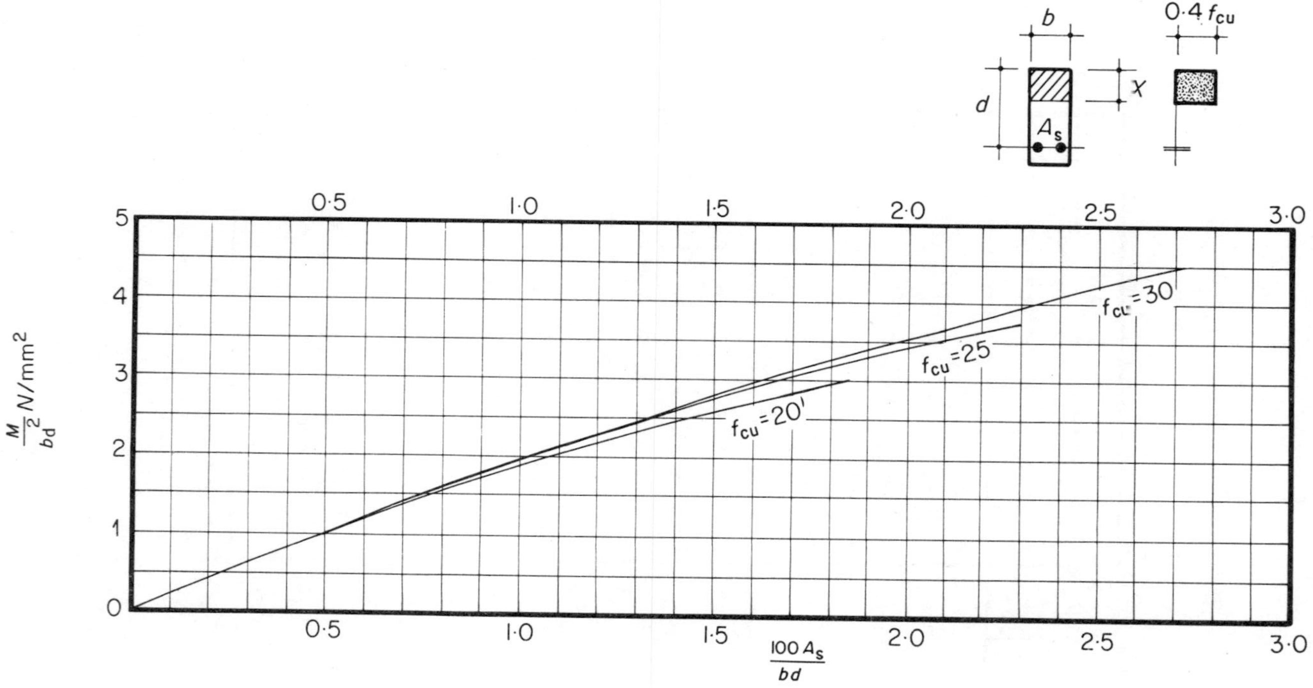

Fig. 5.3 Design chart for singly reinforced beams f_{cu} = 20, 25, 30 N/mm^2, f_y = 250 N/mm^2 $X \not> 0.5d$

thus $$\frac{M}{bd^2} = 0.4 f_{cu} \left(n - \frac{n^2}{2}\right) \qquad \ldots\ldots 5(i)$$

and $$A_s \frac{f_y}{\gamma_m} = 0.4 f_{cu} bnd$$

thus $$\frac{A_s}{bd} = \gamma_m \frac{0.4 f_{cu}}{f_y} n$$

putting $\gamma_m = 1.15$ and multiplying both sides by 100

$$\frac{100 A_s}{bd} = 46 \frac{f_{cu}}{f_y} n \qquad \ldots\ldots 5(ii)$$

Thus for given values of f_{cu} and f_y, a plot of M/bd^2 against $100A_s/bd$ can be made noting that $x = nd$.

The design charts in CP 110: Part 2 for singly reinforced beams are based on the parabolic rectangular stress block and give slight economy in reinforcement when the value of M/bd^2 is high.

5.01 A comparison of design charts for singly reinforced beams using (a) parabolic rectangular stress block, and (b) the simplified rectangular stress block given:

Case (1):

$$f_y = 425 \text{ N/mm}^2$$
$$f_{cu} = 30 \text{ N/mm}^2$$
$$d = 125 \text{ mm}$$
$$b = 1000 \text{ mm}$$
$$\frac{M}{bd^2} = 1.5$$

From chart 3 CP 110 : Part 2 : 1972 for

$$\frac{M}{bd^2} = 1.5 \text{ then } \frac{100A_s}{bd} = 0.425$$

and $$A_s = \frac{0.425 \times 1000 \times 125}{100}$$

$$= 531 \text{ mm}^2$$

From Figure 5.2 $$\frac{100A_s}{bd} = 0.45$$

thus $$A_s = \frac{0.45 \times 1000 \times 125}{100}$$

$$= 562 \text{ mm}^2$$

The difference is only 31 mm^2.

Case (2): As case (1) but with $\frac{M}{bd^2} = 4.0$

From chart 3 CP 110 : Part 2 : 1972 for

$$\frac{M}{bd^2} = 4.0 \text{ then } \frac{100A_s}{bd} = 1.325$$

$$\text{and} \quad A_s = \frac{1.325 \times 1000 \times 125}{100}$$

$$= 1655 \text{ mm}^2$$

From Figure 5.2,

$$\frac{100A_s}{bd} = 1.38$$

$$\text{thus} \quad A_s = \frac{1.38 \times 1000 \times 125}{100}$$

$$= 1725 \text{ mm}^2$$

In this case the difference is 70 mm².

5.02 A reinforced concrete slab is continuous over four equal spans of 6.0 m. Considering the end span only provide a suitable design for the ultimate and serviceability limit states. Assume the following design data:

f_{cu} = 30 N/mm² f_y = 425 N/mm²

partial safety factors for the ultimate limit state are:

γ_m = 1.5 concrete and 1.15 steel
γ_f = 1.4 dead load and 1.6 imposed load.

For the serviceability limit states

γ_m = 1.0 concrete and steel for deflection assessment
γ_m = 1.3 concrete, 1.0 steel for assessment of stresses or crack width at any section.
density of concrete = 2400 kg/m³
characteristic imposed load Q_k = 5 kN/m².

(a) ULTIMATE LIMIT STATE

The starting point for the calculation is an estimate of the self weight of the slab. This could normally be obtained by using the basic span/effective depth ratios given in Chapter 4; but in order to illustrate the effect of adopting a high span to depth ratio on the serviceability calculations a value of 200 mm will be assumed for the slab depth.

$$\text{thus } G_k = 2400 \times 0.2$$

$$= 480 \text{ kg/m}^2$$

$$= 4.71 \text{ kN/m}^2$$

Using the yield line theory with straight yield lines at the support B and in the span AB (Figure 5.4) then with equal reinforcement percentages the design support and span bending moment will be

$$M = \frac{1.4G_k L^2}{11.65} + \frac{1.6Q_k L^2}{11.65}$$

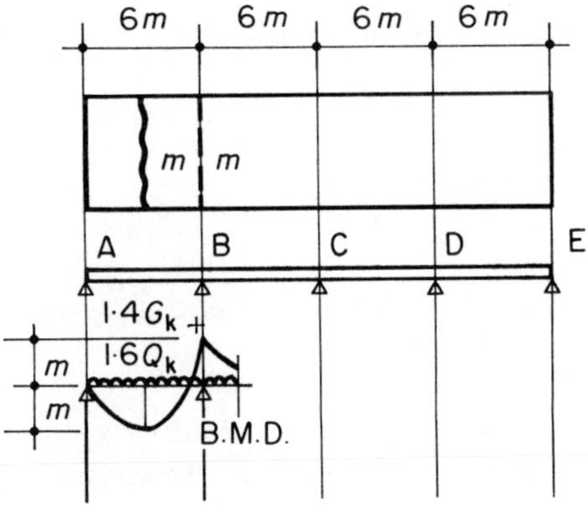

Fig. 5.4

The evaluation of this bending moment coefficient is given in Chapter 2

$$\text{thus } M = 1.4 \times 4.71 \times 6^2/11.65 = 20.4$$

$$= 1.6 \times 5.0 \times 6^2/11.65 = \frac{24.8}{45.2} \text{ kN/m}$$

Using Figure 5.2 and assuming $d = h - 25 = 175$ mm

$$\text{then } \frac{M}{bd^2} = \frac{45.2 \times 10^6}{1000 \times 175^2}$$

$$= 1.475$$

$$\text{thus } \frac{100A_s}{bd} = 0.43$$

$$A_s = \frac{0.43 \times 1000 \times 175}{100}$$

$$= 753 \text{ mm}^2$$

Adopt 12 mm ϕ at 150 mm c/c (754 mm^2/m).

Check shear and bond stresses at support B.

$$\text{the shear force } V = 1.4 \times 4.71 \times 3 = 19.76$$

$$+ 1.6 \times 5.0 \quad \times 3 = 24.00$$

$$+\,45.2/6 \qquad = \frac{7.53}{51.29}\ \text{kN/m}$$

$$\text{shear stress } v = \frac{V}{bd}$$

$$= \frac{51.29 \times 10^3}{1000 \times 175}$$

$$= 0.293\ \text{N/mm}^2$$

This is below the values of v_c given in Chapter 3.

$$\text{the local bond stress } f_{bs} = \frac{V}{\Sigma U_s d}$$

$$\text{and} \qquad \Sigma U_s = \frac{1000}{150} \times 43.6$$

$$= 290\ \text{mm}$$

$$\text{thus} \qquad f_{bs} = \frac{51.29 \times 10^3}{290 \times 175}$$

$$= 1.01\ \text{N/mm}^2$$

This is within the ultimate local bond stress values given in Chapter 3. Thus it would appear that a 200 mm deep slab with 12 mm ϕ bars at 150 mm c/c is adequate for the ultimate limit state. The reinforcement percentage

$$P = \frac{754}{1000 \times 175} \times 100$$

$$= 0.426$$

The span effective depth ratio is 6/0.175 = 34.3 and thus the design is suspect in terms of serviceability.

(b) SERVICEABILITY LIMIT STATE – CRACKING

The yield line theory used for the ultimate limit state gives equal moments at the support and in the span for the same reinforcement percentages. For the estimation of the maximum bending moment at support B, at which a crack width check will be made, bending moment coefficients based on an elastic analysis will be used. It is normal practice to base this analysis on a constant flexural stiffness derived from the entire member cross section ignoring the presence of reinforcement as indicated in Chapter 1. Although in practice the slab will, in part, be cracked at the service load, a quantification of the effect of this on the distribution of bending moments is not practical. Thus it is proposed to use the bending moment coefficients given in Figure 1.13. With $\gamma_f = 1.0$ the bending moment at support B (Figure 5.5) is given by

$$M_{\text{elastic}} = 0.121 Q_k L^2 + 0.107 G_k L^2$$

$$= 0.121 \times 5.0 \times 6^2 = 21.7$$

$$= 0.107 \times 4.71 \times 6^2 = \frac{18.1}{39.8} \text{ kNm/m}$$

The value of γ_m is taken as 1.3 and thus the design strength

$$= \frac{f_{cu}}{1.3} = \frac{30}{1.3}$$

$$= 23 \text{ N/mm}^2$$

From Table 4.1 the elastic modulus for the concrete is 25.6 kN/mm^2.

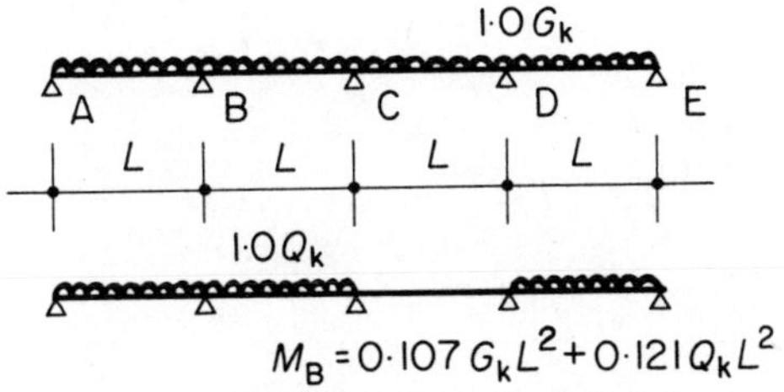

Fig. 5.5

In assessing strains the modulus of elasticity of the concrete is taken as one half of the instantaneous value = 0.5 x 25.6 x 12.8 kN/mm^2. Assuming the modulus of elasticity of the steel is constant at 200 kN/mm^2 the modular ratio is given by:

$$\alpha e = \frac{200}{12.8} = 15.6 \quad \text{(say 15)}$$

From Table 1.1 (Chapter 1) for $\alpha e = 15$ and $p = 0.426$ the neutral axis factor n is in the order of 0.3

$$\text{thus } x = 0.3 \times 175$$

$$= 52.5 \text{ mm}$$

$$\text{and the lever arm } z = d - \frac{x}{3}$$

$$= 175 - 17.5$$

$$= 157.5 \text{ mm}$$

$$\text{the steel stress } f_s = \frac{M_{\text{elastic}}}{A_s z}$$

$$= \frac{39.8 \times 10^6}{754 \times 157.5}$$

$$= 335 \text{ N/mm}^2$$

$$\text{the corresponding steel strain} = \frac{335}{200 \times 10^3}$$

$$= 0.001\,675$$

This ignores the stiffening effect of the concrete in the tension zone but is still just below the limiting value of $0.8\, f_s/E_s = 0.0017$

$$\text{the concrete stress } f_c = \frac{2M}{bxZ}$$

$$= \frac{2 \times 39.8 \times 10^6}{1000 \times 52.5 \times 157.5}$$

$$= 9.6 \text{ N/mm}^2$$

Using the crack width formula in C & CA Research Report No. 18 (see Chapter 4) then the crack width is given by

$$w = Kce$$

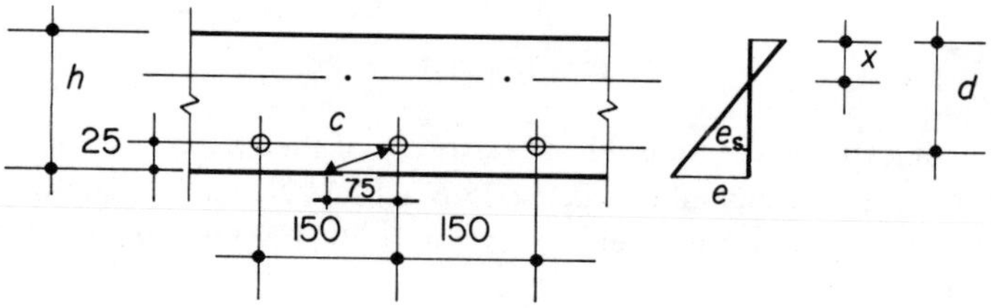

Fig. 5.6

where K = 3.3 for deformed bars and from Figure 5.6,

$$e = e_s \frac{h - x}{d - x}$$

$$= 0.001\,675 \times \frac{200 - 52.5}{175 - 52.5}$$

$$= 2.02 \times 10^{-3}$$

Considering the width of crack midway between two 12 mm bars

$$c = (75^2 + 25^2)^{\frac{1}{2}} - 6$$

$$= 73 \text{ mm}$$

$$\text{thus} \quad w = 3.3 \times 73 \times 2.02 \times 10^{-3}$$

$$= 0.486 \text{ mm}$$

This is a high value and ignores the stiffening effect of the concrete in the tension zone. To allow for this the strain calculated from elastic theory can be reduced by an amount

$$\frac{4bd}{A_s} 10^{-6} \quad \text{(see Chapter 4)}$$

$$\text{thus net strain} = 2.02 \times 10^{-3} - \frac{4 \times 1000 \times 175 \times 10^{-6}}{754}$$

$$= 2.02 \times 10^{-3} - 9.28 \times 10^{-4}$$

$$= 0.001\,092$$

$$\text{and} \quad w = 0.001\,092 \times 73 \times 3.3$$

$$= 0.263 \text{ m}$$

If K is taken as 2.3 (draft CP 110) this value reduces to 0.18 mm, see Chapter 4. This result will be compared with the value obtained from the formula for assessing crack widths given in Appendix A of CP 110.

$$\text{design surface crack width } w = \frac{3A_{cr} \times \epsilon_m}{1 + 2\left[\dfrac{A_{cr} - C_{min}}{h - x}\right]}$$

The notation is defined in Chapter 4 and in this case

$$A^1 - x = h - x$$

$$\text{and} \quad \epsilon_m = e_1 - \frac{1.2bth(A^1 - x)}{A_s(h - x)f_y} \times 10^{-3}$$

$$= e_1 - \frac{1.2 \times 1000 \times 200 \times 10^{-3}}{754 \times 425}$$

$$= e_1 -\!- 0.000\,75$$

From the previous calculation $e_1 = 0.002\,02$ and thus $\epsilon_m = 0.001\,27$. From Figure 5.7, $A_{cr} = 73$ mm, $C_{min} = 19$ mm and $h - x = 147.5$ mm

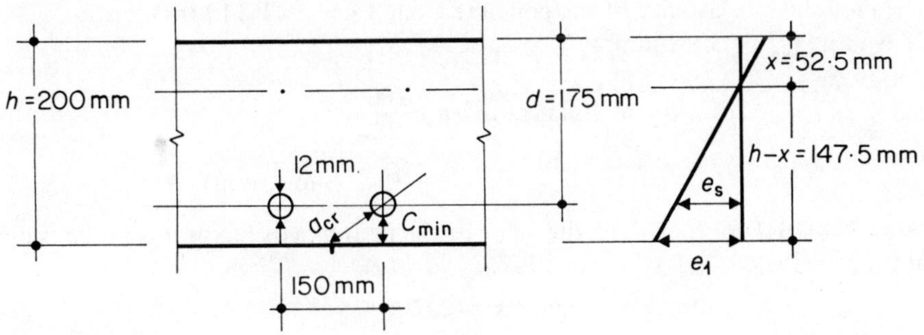

Fig. 5.7

$$\text{thus} \quad w = \frac{3 \times 73 \times 1.27 \times 10^{-3}}{1 + 2\left[\dfrac{73 - 19}{147.5}\right]}$$

$$= 0.16 \text{ mm}$$

(c) SERVICEABILITY LIMIT STATE – DEFLECTION

For the calculation of the deflection it will be necessary to estimate the maximum moment in span AB and thus the loading arrangement will differ from that used for the crack width calculation. Using the loading arrangement shown in Figure 5.8 the midspan moment approximates to

$$M_{\text{elastic}} = \frac{G_k L^2}{13} + \frac{Q_k L^2}{10}$$

$$= 4.71 \times \frac{6^2}{13} + 5.0 \times \frac{6^2}{10}$$

$$= 13.05 + 17.5$$

$$= 30.53 \text{ kN/m}$$

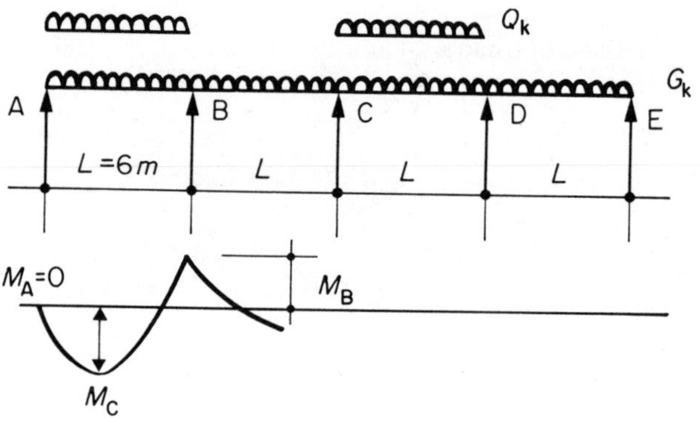

Fig. 5.8

The modulus of elasticity of the concrete from Table 1 CP 110 is taken as 28 N/mm^2 (f_{cu} = 30 N/mm^2)

$$\text{thus the modular ratio } \alpha e = \frac{200}{28}$$

$$= 7.15 \quad \text{(short term)}$$

From Table 1.1 for αe = 7.15 the value of the neutral axis factor n is in the order of 0.22

$$\text{thus} \quad x = 0.22 \times 175$$

$$= 35 \text{ mm}$$

Using the equilibrium of moments equation developed in Chapter 4 (equation 4(x)), then

$$M_{\text{elastic}} = 30.53 \times 10^6 = \frac{pbd}{100} f_s\left(d - \frac{x}{3}\right) + \frac{bh}{3}\,\frac{(h-x)^2}{d-x}$$

and for $d = 175$ mm, $h = 200$ mm and $p = 0.43$

$$30.53 = 0.123 f_s + 12.95$$

$$\text{thus}\quad f_s = \frac{17.58}{0.123}$$

$$= 143 \text{ N/mm}^2$$

$$\text{curvature } \phi = \frac{1}{\gamma_b} = \frac{f_s}{E_s(d-x)}$$

$$= \frac{143}{200 \times 10^3 \times 140}$$

$$= 5.1 \times 10^{-6}$$

The deflection can be expressed as $K\phi L^2$ where K for a continuous beam is given by

$$K = 0.104\left(1 - \frac{B}{10}\right)$$

$$\text{and}\quad B = \frac{M_A + M_B}{M_c} \quad \text{(see Figure 5.8)}$$

In this case $M_A = 0$ and

$$M_B = 0.107 G_k L^2 + 0.054 Q_k L^2$$

$$= 0.107 \times 4.71 \times 6^2 + 0.054 \times 5.0 \times 6^2$$

$$= 18.15 + 9.72$$

$$= 27.87 \text{ kN/m}$$

$$\text{thus}\quad B = \frac{27.87 + 0}{30.53}$$

$$= 0.915$$

$$\text{thus } 1 - B/10 = 0.9085$$

$$\text{and the deflection} = 0.104 \times 0.9085 \times 5.1 \times 10^{-6} \times 6^2 \times 10^6$$

$$= 17.35 \text{ mm}$$

This represents the instantaneous deflection under the characteristic dead and imposed loadings. As indicated in Chapter 4 a precise evaluation of the long term deflection is virtually impossible due to the large number of variables involved. In this example the influence of shrinkage will be ignored as the reinforcement percentage is low and will provide only a moderate amount of restraint to the movement of the concrete round it. If it is assumed that the duration of the characteristic imposed load is infinite and the age at which it is applied is, say, 28 days, the creep coefficient ϕ, taken from Table A.1 in the Handbook on the Unified Code for Structural Concrete[34] is 2.75 and the tensile stress in the concrete reduces to 0.55 N/mm^2. Thus the effective modulus is given by

$$E_{\text{eff}} = E_c \frac{1}{1+\phi}$$

where E_c is the short term modulus and taken as 28 kN/mm^2

$$\text{thus} \quad E_{\text{eff}} = \frac{28}{1+2.75}$$

$$= 7.47 \text{ kN/mm}^2$$

$$\text{the modular ratio} = \frac{200}{7.47}$$

$$= 26.8$$

From Table 1.1 for $\alpha e = 26.8$ the neutral axis factor n approximates to 0.37.

$$\text{hence} \quad x = 0.37 \times 175$$

$$= 64.8 \text{ mm}$$

$$d - x/3 = 175 - 21.6$$

$$= 153.4 \text{ m}$$

$$\text{and} \quad h - x = 200 - 64.8$$

$$= 135.2 \text{ mm}$$

Using the equilibrium of moments equation with the tensile stress in the concrete as 0.55 N/mm^2 in the long term

$$\frac{pbd}{100} f_s \left(d - \frac{x}{3}\right) = 0.0043 \times 10\,000 \times 175 \times 153.4 \times f_s$$

$$= 0.1154 \, f_s$$

$$\text{and} \quad 0.55 \frac{bh}{3} \frac{(h-x)^2}{d-x} = \frac{1000 \times 200 \times 135.2^2 \times 0.55}{3 \times 110.2}$$

$$= 6.09$$

$$\text{thus} \quad M_{\text{elastic}} = 30.53 \times 10^6$$

$$= 0.1154 f_s + 6.09$$

$$\text{and} \quad f_s = 212 \text{ N/mm}^2$$

$$\text{The curvature } \phi = \frac{f_s}{E_s(d-x)}$$

$$= \frac{212}{200 \times 10^3 \times 110.2}$$

$$= 9.62 \times 10^{-6}$$

$$\text{Thus deflection} = K\phi L^2$$

$$= 17.35 \times \frac{9.62}{5.1}$$

$$= 32.8\,\text{mm}$$

$$\text{The ratio } \frac{\text{span}}{\text{deflection}} = \frac{6000}{17.35} = 346 \text{ short term}$$

$$\text{and } \frac{6000}{32.8} = 189 \text{ long term}$$

The long term deflection calculation has been based on the assumption that the characteristic imposed loading is permanent and applied in full when the concrete has reached an age of 28 days. Thus the ratios of 189 and 346 represent two extremes and the actual long term deflection will be somewhere between these values.

5.03 A reinforced concrete slab (Figure 5.9), built in on three sides and free on the fourth is subjected to an equivalent fluid pressure. Provide an outline design for the ultimate limit state given the following data:

density of fluid = 480 kg/m^3
partial safety factor γ_f = 1.6

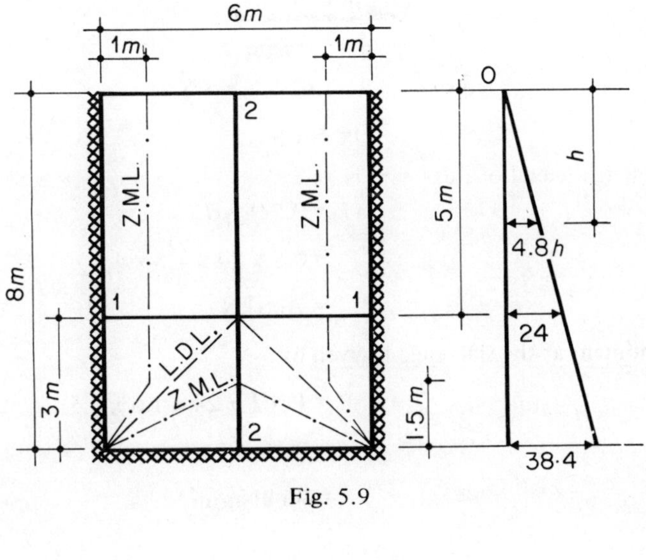

Fig. 5.9

characteristic strength of concrete = 30 N/mm^2
characteristic strength of reinforcement = 425 N/mm^2

The fluid pressure varies from zero to a maximum at a depth of 8 m. The density of fluid is 480 kg/m^2 which is approximately 4.8 kN/m^3. At a depth h the hydrostatic pressure is $4.8h$. The slab will be analysed using the Hillerborg strip method, the first step being to choose the positions of the load dispersion and zero moment lines. This choice should be made bearing in mind the elastic distribution of moments at service load otherwise crack widths may be excessive. Considering strip 1–1 as a beam built in at both ends, an elastic analysis would give the point of contraflexure at a distance of 0.21 x 6 = 1.26 m from the supports. Thus for the ultimate limit state the position of the zero moment line from the support should not be too far from this position, and a value of 1.0 m has been adopted.

Strip 1–1
At a depth of 5.0 m the hydrostatic pressure is 4.8 x 5 = 24 kN/m^2. If the width of the strip is taken as 1 m the pressure will vary between 4 x 4.8 = 19.2 and 24 kN/m^2 over a metre width of strip. Thus the loading is as shown in Figure 5.10. Between the Z.M.L.s there is a simply supported span of L = 4 m. For simplicity a value of 24 kN/m^2 and a metre width of strip have been assumed

$$\text{the bending moment } M = 0.125\,Q_k \gamma_f L^2$$
$$= 0.125 \times 24 \times 1.6 \times 4^2$$
$$= 76.8 \text{ kNm/m}$$

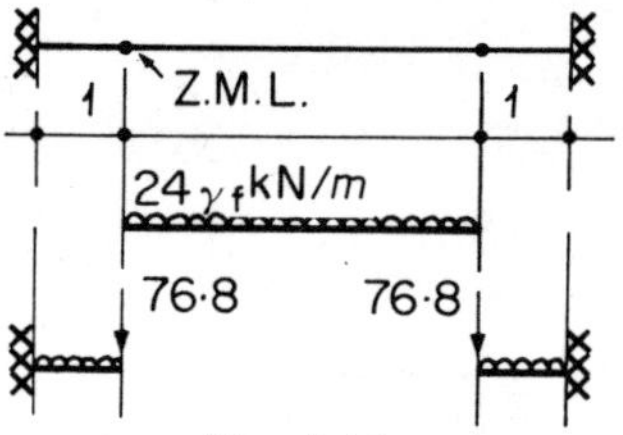

Fig. 5.10

The reaction at the end of this beam is

$$V = 0.5\,Q_k \gamma_f L$$
$$= 0.5 \times 24 \times 1.6 \times 4$$
$$= 76.8 \text{ kN}$$

Thus the moment at the slab edge is given by

$$M = V \times 1 + 24 \times 1.6 \times 0.5$$
$$= 76.8 + 19.2$$
$$= 96 \text{ kNm/m}$$

Strip 2–2

The loading on strip 2–2 is shown on Figure 5.11. In order to maintain equilibrium a reaction R_2 is required at the free edge (for which additional reinforcement must be provided) and considering the partially loaded beam between R_1 and R_2 (γ_f omitted)

$$R_2 = 24 \times 1.5 \times \frac{0.75}{6.5} + 3.6 \times 1.5 \times \frac{0.5}{6.5}$$

$$= 4.57\ \text{kN}$$

$$R_1 = 24 \times 1.5 \times \frac{5.75}{6.5} + 3.6 \times 1.5 \times \frac{6.0}{6.5}$$

$$= 36.83\ \text{kN}$$

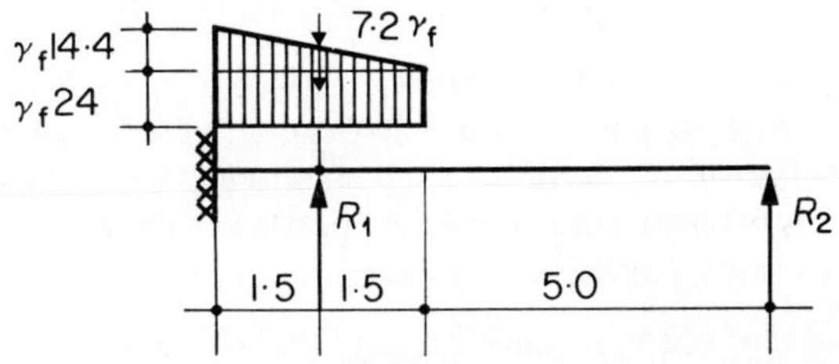

Fig. 5.11

The bending moments at various positions along the beam can now be evaluated. As with strip 1–1 the metre width strip is assumed to be uniformly loaded

$$M_{5.0} = R_2 \times 5.0$$

$$= 4.57 \times 5 \times 1.6$$

$$= 36.6\ \text{kNm/m}$$

$$M_{8.0} = 36.83 \times 1.5 \times 1.6 = 88.2$$

$$24 \times 1.5 \times 0.75 \times 1.6 = 43.2$$

$$7.2 \times 1.5 \times 0.75 \times 1.6 = 12.96$$

$$3.6 \times 1.5 \times 0.5 \times 1.6 = 4.32$$

$$148.68\ \text{kNm}$$

Considering the moment of 148.68 kNm the reinforcement area will be calculated using Figure 5.2; assume $h = 250$ mm.

$$d = h - 50 \text{ (say)}$$

$$= 200\ \text{mm}$$

$$f_{cu} = 30\ \text{N/mm}^2$$

$$f_y = 425 \text{ N/mm}^2$$

$$\frac{M}{bd^2} = \frac{148.68 \times 10^6}{1000 \times 200^2}$$

$$= 3.72$$

From Figure 5.2 for $f_{cu} = 30 \text{ N/mm}^2$

$$\frac{100A_s}{bd} = 1.25$$

$$\text{thus } A_s = \frac{1000 \times 200 \times 1.25}{100}$$

$$= 2500 \text{ mm}^2/\text{m}$$

Adopt 20 mm ϕ at 125 mm c/c (2510).

5.04 A warehouse floor is to be designed as a reinforced concrete flat slab supported by columns on a grid of 6 m x 6 m. Columns are to be detailed such that the moment transferred from the floor is minimized and lateral stability is provided by means of shear walls. Consider the ultimate and serviceability limit states using the data as below:

$f_{cu} = 30 \text{ N/mm}^2$, $f_y = 425 \text{ N/mm}^2$, $Q_k = 15 \text{ kN/m}^2$, partial safety factors as Example 5.02.

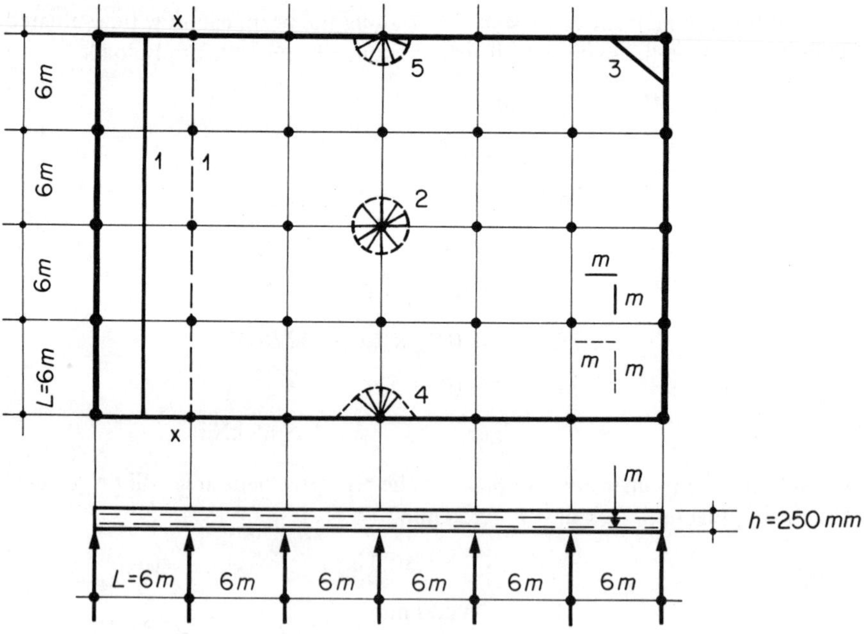

Fig. 5.12

(a) THE ULTIMATE LIMIT STATE

For the analysis of the flat slab the yield line theory will be adopted with a trial depth of 250 mm. The design loadings are as follows

$$G_k = 0.25 \times 2400$$
$$= 600\,\text{kg/m}^3$$
$$= 5.90\,\text{kN/m}^2$$

$$\text{thus design dead load} = 1.4 \times 5.9$$
$$= 8.26\,\text{kN/m}^2 \text{ (maximum)}$$

$$\text{design imposed load} = 1.6 \times Q_k$$
$$= 1.6 \times 15$$
$$= 24\,\text{kN/m}^2$$

In the design of a warehouse floor it is generally necessary to examine the effects of chequerboard loading and thus with high imposed loadings large negative moments can be developed in the span between the column lines. For simplicity of design and detailing a constant mesh top and bottom will be assumed initially. The analysis of the slab can now be based on isotropic conditions with $m = im$, as $i = 1.0$. The general arrangement of the slab and possible failure modes are shown in Figure 5.12. With regard to failure mode 1, the assumption of a constant mesh top and bottom precludes the development negative yield lines in the span for loading as indicated on Figure 5.13.

Failure mode 1

Straight yield lines which for a constant mesh in the top and bottom of the slab will give for the end bay

$$m = \frac{\gamma_f Q_k L^2}{11.65} + \frac{\gamma_f G_k L^2}{11.65}$$

$$= \frac{6^2}{11.65}(24 + 8.26)$$

$$= 99.9\,\text{kNm/m}$$

Failure mode 2

A full circular fan with the column load P approximating to $32.26 \times 6^2 = 1162\,\text{kN}$ (continuity factor ignored);

$$\text{thus} \quad m = \frac{P}{2\pi(1 + i)} = \frac{P}{4\pi} \; (i = 1.0)$$

$$= \frac{1162}{4\pi}$$

$$= 92.5\,\text{kNm/m}$$

Failure mode 3

With $i = 1.0$ the failure mode is a single yield line with

$$P = 2m$$

where P approximates to $0.25 \times 1162 = 290.5$ kN

$$\text{thus} \quad m = \frac{P}{2}$$

$$= \frac{290.5}{2}$$

$$= 145.25 \text{ kNm}$$

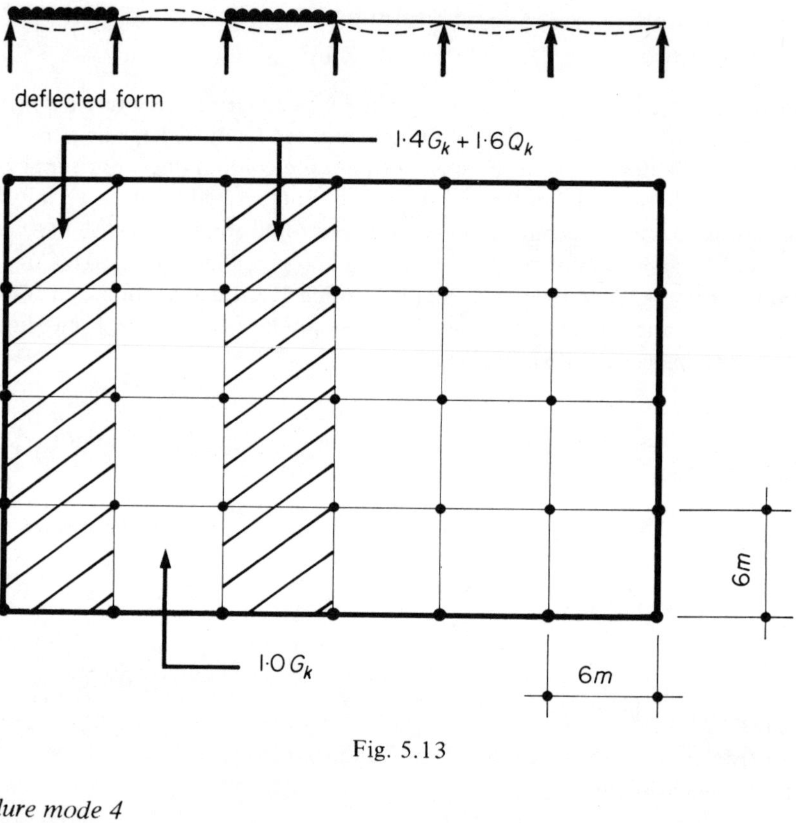

Fig. 5.13

Failure mode 4

With $i = 1.0$ $\qquad P = 5.14m$

where P approximates to $0.5 \times 1162 = 581$ kN

$$\text{thus} \quad m = \frac{P}{5.14}$$

$$= \frac{581}{5.14}$$

$$= 113 \text{ kNm/m}$$

Note if a semicircular fan were assumed

$$m = \frac{581}{2\pi}$$

$$= 92.5 \text{ kNm/m}$$

As mentioned in Chapter 3, this mode of failure is suspect as radial moments are required at the slab edge. From the four modes investigated the maximum moment is 145.25 kNm/m. The reinforcement required for this moment is calculated using Figure 5.2.

$$d = h - 38 = 250 - 38$$

$$= 212 \text{ mm}$$

$$\frac{M}{bd^2} = \frac{145.25 \times 10^6}{1000 \times 212^2}$$

$$= 3.24$$

Thus for $f_{cu} = 30$ N/mm^2 and $f_y = 425$ N/mm^2,

$$\frac{100A_s}{bd} = 1.05$$

$$A_s = \frac{1000 \times 212 \times 1.05}{100}$$

$$= 2230 \text{ mm}^2\text{/m}$$

Adopt 16 mm ϕ at 90 mm c/c (2235). This reinforcement would be provided both ways, top and bottom in the four corner bays and then reduced in accordance with the critical failure mode for the other bays. It is also necessary to investigate shear, and failure could occur along one of the column lines, say section x–x (Figure 5.12). Considering a metre width and a distribution of moments as mode 1,

$$V = 32.26 \times 3 + \frac{99.9}{6}$$

$$= 96.9 + 16.7$$

$$= 113.6 \text{ kN/m}$$

$$\text{shear stress } v = \frac{V}{bd}$$

$$= \frac{113.6 \times 10^3}{1000 \times 212}$$

$$= 0.536 \text{ N/mm}^2$$

For 100, $A_s/bd = 0.5$, the value of v_c from Table 5 CP 110, with $f_{cu} = 30$ N/mm², is 0.55, thus $v < v_c$. The value of 100 A_s/bd for the corner bays is 1.05 and for the other bays it will not be less than 0.5. A more critical condition is that of punching shear round a column head. For an internal column the load is 1162 kN to which a continuity factor of, say, 1.15 will be added

$$\text{thus shear} = 1162 \times 1.15$$

$$= 1335 \text{ kNm}$$

The critical section is taken on a perimeter $1.5h$ from the boundary of the loaded area, see Figure 5.14. Assuming a column 400 mm x 400 mm the critical perimeter is $= 4 \times 400 + 2\pi \times 1.5h = 3955$ mm

$$\text{thus} \quad v = \frac{1335 \times 10^3}{3955 \times 212}$$

$$= 1.59 \text{ N/mm}^2$$

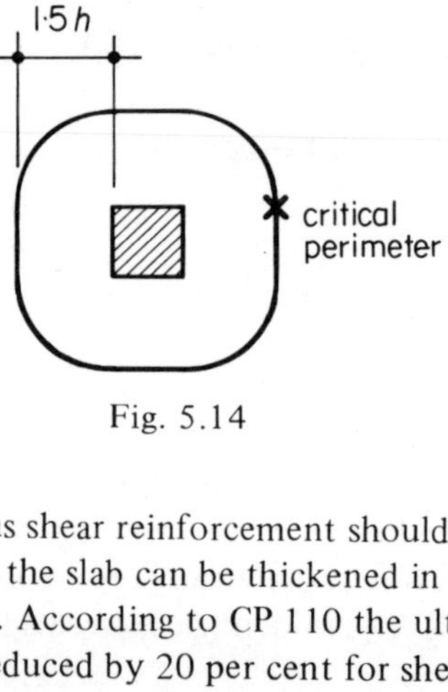

Fig. 5.14

In this case $v > v_c$ and thus shear reinforcement should be provided, or alternatively the thickness of the slab can be thickened in the region of the column to form a drop panel. According to CP 110 the ultimate shear stresses given by Table 5 should be reduced by 20 per cent for shear in flat slabs. Assuming a value of 100 A_s/bd approximating to 1.0 then the effective depth of slab required can be obtained as follows

$$\text{for } f_{cu} = 30 \text{ N/mm}^2 \quad v_c = 0.8 \times 0.7 = 0.56 \text{ N/mm}^2$$

$$\text{assume } h \sim 1.1d$$

$$\text{the critical perimeter} = 1600 + 2 \times \pi \times 1.5 \times 1.1d$$

$$= 1600 + 10.35d$$

$$\text{thus } d = \frac{V}{0.56\,(1600 + 10.35d)}$$

$$1335 \times 10^3 = 0.56d(1600 + 10.35d)$$

$$= 896d + 5.8d^2$$

$$d^2 + 154.5d - 230 \times 10^3 = 0$$

$$d = -\frac{154.5}{2} \pm \tfrac{1}{2}(154.5^2 + 4 \times 230 \times 10^3)^{\frac{1}{2}}$$

$$= 409 \text{ mm}$$

Consider the arrangement as shown in Figure 5.15. Providing h_1 is not less than 409 x 1.1, say 450 mm, section 1–1 is adequate for shear. If the drop panel is 2.0 m square another critical section is 2–2. The shear V at 2–2 approximates to $1.15 \times 32.26\,(6^2 - 2.75^2) = 1056$ kN.

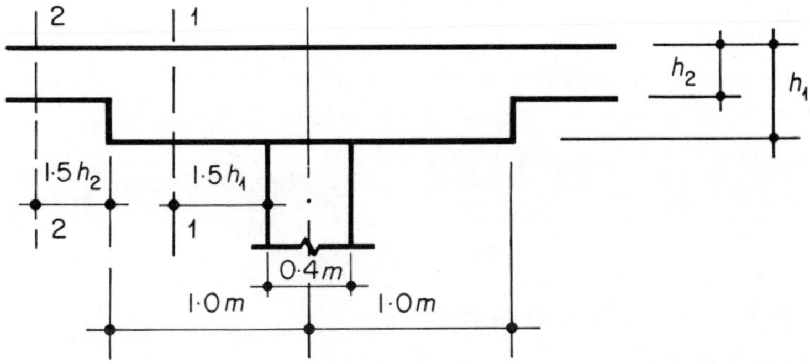

Fig. 5.15

The critical perimeter is

$$= 4 \times 2000 + 2 \times \pi \times 1.5 \times 250$$

$$= 8000 + 2355$$

$$= 10\,355 \text{ mm}$$

$$\text{thus } v = \frac{1056 \times 10^3}{10\,355 \times 212}$$

$$= 0.48 \text{ N/mm}^2$$

This is less than the limiting value of 0.56 N/mm^2.

(b) SERVICEABILITY LIMIT STATE – CRACKING

As an approximation it will be assumed that the moments at the service load are

70 per cent of those relating to the ultimate limit state. Considering the maximum moment of 145.25 the reinforcement provided is 2235 mm²/m. Thus the reinforcement percentage

$$P = \frac{A_s}{bd}$$

$$= \frac{2235}{1000 \times 212} \times 100$$

$$= 1.05$$

The modulus of elasticity for the concrete ($\gamma_m = 1.3$) is 25.6 N/mm² (see Example 5.02) and thus

$$\alpha e = \frac{200}{12.8}$$

$$= 15.6 \quad \text{(say 15)}$$

From Table 1.1 for $\alpha e = 15$ and $P = 1.05$, the neutral axis factor n approximates to 0.42

$$x = nd = 0.42 \times 212$$

$$= 89.04\,\text{m}$$

$$\text{and} \quad Z = d - \frac{x}{3} = 212 - 29.68$$

$$= 182.32\,\text{mm}$$

$$\text{The steel stress } f_s = \frac{M_{\text{elastic}}}{A_s \times Z}$$

$$= \frac{0.7 \times 145.25 \times 10^6}{2335 \times 182.32}$$

$$\text{thus} \quad f_s = 249.5\,\text{N/mm}^2$$

The corresponding strain is

$$= \frac{249.5}{200 \times 10^3}$$

$$= 0.001\,25$$

This ignores the stiffening effect of the concrete in the tension zone, and by inspection for a bar spacing of 90 mm the crack widths will not be excessive.

(c) SERVICEABILITY LIMIT STATE – DEFLECTION

$$\text{the ratio, span/effective depth} = \frac{6000}{212}$$

$$= 28.3$$

As the slab is reinforced top and bottom both ways, it is considered that this ratio is reasonable and further calculations are not required.

5.05 A reinforced concrete bridge pier is fixed at the base, free at the top and subjected to design ultimate loads as indicated in Figure 5.16. Using the following data determine the reinforcement required for the ultimate limit state.

$f_{cu} = 30\ \text{N/mm}^2,\ f_y = 425\ \text{N/mm}^2$

$\left.\begin{array}{l} N = 2400\ \text{kN} \\ H = 200\ \text{kN} \end{array}\right\} \gamma_f$ values included

$h = 600\ \text{mm}$

$b = 1200\ \text{mm}$

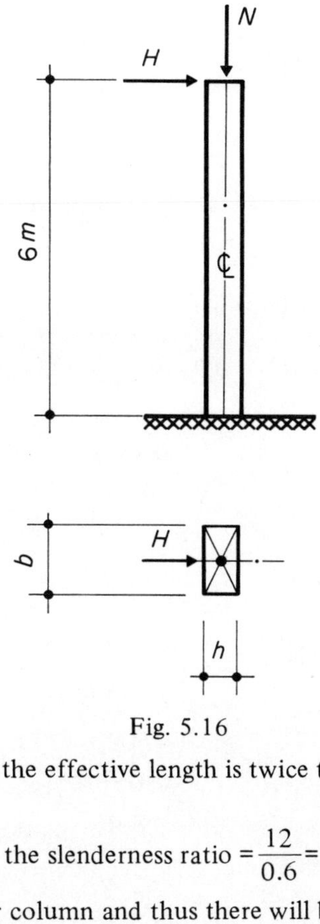

Fig. 5.16

For a cantilever column the effective length is twice the actual height = 2 x 6 = 12 m = l_e

$$\text{the slenderness ratio} = \frac{12}{0.6} = 20$$

This represents a slender column and thus there will be an amplification of moment which can be evaluated using the expression developed in Section 3(f). The total moment is given by

$$M_t = M_1 + \frac{Nh}{1750}\left(\frac{l_e}{h}\right)^2 \left(1 - 0.0035\frac{l_e}{h}\right)k$$

Initially it will be assumed that the factor k, which represents the influence of axial load on reducing the curvature, equals unity.

$$M_1 = H \times 6 = 200 \times 6$$
$$= 1200 \text{ kNm}$$

$$\frac{Nh}{1750}\left(\frac{l_e}{h}\right)^2\left(1 - 0.0035\frac{l_e}{h}\right) = \frac{2400 \times 0.6}{1750}\left(\frac{12}{0.6}\right)^2 (1 - 0.0035 \times 20)$$
$$= 306 \text{ kNm}$$

$$\text{thus} \quad M_t = 1200 + 306$$
$$= 1506 \text{ kNm}$$

The influence of axial load will now be investigated. The factor k is given by

$$k = \frac{N_u z - N}{N_u z - N_{BAL}}$$

(notation given in Chapter 3).

In this example the value of N is small and will be less than N_{BAL} and thus

$$N_u z - N > N_u z - N_{BAL}$$

Therefore the value of k may be taken as unity.

The reinforcement area is most conveniently obtained by using chart 84 in part 2 of CP 110 : 1972. For $d/h = 0.9$ say, $f_y = 425 \text{ N/mm}^2$ and $f_{cu} = 30 \text{ N/mm}^2$.

$$\frac{N}{bh} = \frac{2400 \times 10^3}{1200 \times 600}$$
$$= 3.33$$
$$\frac{M}{bh^2} = \frac{1506 \times 10^6}{1200 \times 600^2}$$
$$= 4.71$$

$$\text{thus} \quad \frac{100A_{sc}}{bh} = 2.4$$

$$\text{and} \quad A_{sc} = \frac{2.4 \times 600 \times 1200}{100}$$
$$= 1728 \text{ mm}^2$$

Adopt six No. 20 mm ϕ (1890).

5.06 Given the following data determine the maximum torsional moment appropriate to the ultimate limit state and the reinforcement required to resist this torsion.

$$M = 448 \text{ kNm}$$

$$V = 280\ \text{kN}$$

$$h = 0.575\ \text{m}$$

$$b = 0.3\ \text{m}$$

$$d = 0.5\ \text{m}$$

$$A_s = 2810\ \text{mm}^2$$

$$f_{cu} = 30\ \text{N/mm}^2$$

$$f_{yv} = f_{yl} = 425\ \text{N/mm}^2$$

The above values of M and V relate to ultimate conditions and the shear stress due to V is given by

$$v = \frac{V}{bd}$$

$$= \frac{280 \times 10^3}{500 \times 300}$$

$$= 1.86\ \text{N/mm}^2$$

The sum of the ultimate shear stresses due to V and the torsional moment T should not exceed the value v_{tu} given in the table in CP 110. For $f_{cu} = 30\ \text{N/mm}^2$, $v_{tu} = 4.1\ \text{N/mm}^2$

$$\text{hence } 4.1 = v + v_t$$

$$v_t = 4.1 - 1.86$$

$$= 2.24\ \text{N/mm}^2$$

Nor should v_t exceed $v_{tu} \times \dfrac{y_1}{550}$

Referring to Figure 5.17, $y_1 = 450$ mm

$$\text{thus } v_{tu} \times \frac{y_1}{550} = 4.1 \times \frac{450}{550}$$

$$= 3.35\ \text{N/mm}^2\ \text{O.K.}$$

The torsional shear stress for a rectangular section is given by

$$v_t = \frac{2T}{\left[h_{min}^2 h_{max} - \dfrac{h_{min}}{3}\right]}$$

In this case

$$h_{min} = 0.300\ \text{m}$$

$$h_{max} = 0.575\ \text{m}$$

$$\text{thus } h_{min}^2 \left[h_{max} - \frac{h_{min}}{3}\right] = 0.3^2 \left[0.575 - \frac{0.3}{3}\right]$$

$$= 0.09 \times 0.475$$

$$= 0.04275$$

hence $T = 0.5 \times 2.24 \times 0.042\,75 \times 10^3$

$$= 47.88 \text{ kNm}$$

Try two 10 mm ϕ legs then $A_{sv} = 157 \text{ mm}^2$

$$\frac{A_{sv}}{s_v} = \frac{T}{0.8 \times x_1 \times y_1 \times (0.87 f_{yv})}$$

$$\text{and } s_v = \frac{157 \times 0.8 \times 225 \times 450 \times 0.87 \times 425}{47.88 \times 106}$$

$$= 98.2 \text{ say } 100 \text{ mm c/c}$$

$$A_{sl} = \frac{A_{sv}}{s_v}\left(\frac{f_{yv}}{f_{yl}}\right)(x_1 + y_1)$$

$$= \frac{157}{98.2} \times 1 \times (225 + 425)$$

$$= 1039 \text{ mm}^2$$

Adopt four 20 mm ϕ (1260).

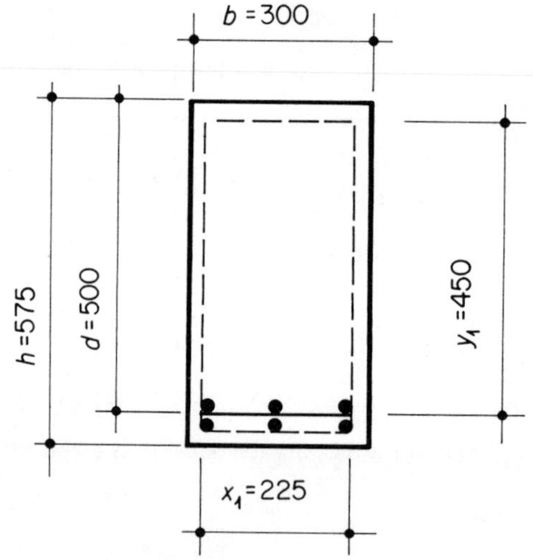

Fig. 5.17

It should be noted that the reinforcement A_{sv} and A_{sl} is additional to that required for shear and bending.

5.07 A reinforced concrete double cantilever beam has dimensions as indicated in Figure 5.18. Given the following data estimate the reinforcement requirements for the ultimate limit states for bending and shear.

Q_k = 21 kN/m, G_k = 46 kN/m, f_{cu} = 30 N/mm^2, f_s = 425 N/mm^2, beams at 7.5 m centres, slab thickness 225 mm, γ_f = 1.4 for G_k and 1.6 for Q_k, γ_m = 1.5 concrete and 1.15 steel.

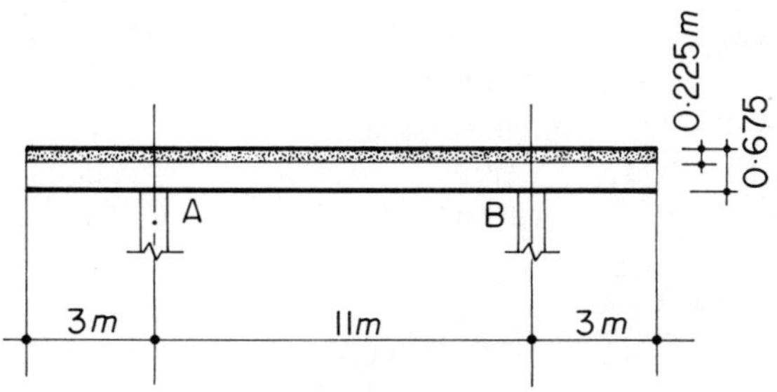

Fig. 5.18

The maximum moment for the cantilever is given by

$$M = 1.4 \times 46 \times 3^2 \times 0.5 + 1.6 \times 21 \times 3^2 \times 0.5$$
$$= 289.8 + 151.2$$
$$= 441 \text{ kNm}$$

The maximum positive moment is obtained by loading AB with $1.4G_k + 1.6Q_k$ and the cantilevers with $1.0G_k$. In this case

$$\text{cantilever moment} = 1.0 \times 46 \times 3^2 \times 0.5$$
$$207 \text{ kNm}$$

The simple beam moment for AB is

$$= (1.4 \times 46 + 1.6 \times 21) \times 11^2 \times 0.125$$
$$= 1482 \text{ kNm}$$

$$\text{thus span moment} = 1482 - 207$$
$$= 1275 \text{ kNm}$$

The effective sections at the supports and in the span are shown in Figures 5.19 and 5.20. The support section is subjected to a moment of 441 kN and the reinforcement is calculated from Figure 5.2.

$$d = 600 \text{ mm}$$

$$\frac{M}{bd^2} = \frac{441 \times 10^6}{300 \times 600^2}$$
$$= 4.08$$

$$\text{thus} \quad \frac{100A_s}{bd} = 1.42 \text{ for } f_{cu} = 30 \text{ N/mm}^2$$

$$A_s = \frac{300}{100} \times 600 \times 1.42$$

$$= 2556 \text{ mm}^2$$

Adopt four 25 mm ϕ and two 20 mm ϕ ($1960 + 628 = 2588$ mm^2).
Considering the span section it is necessary to determine the effective width B. According to CP 110 this may be taken as:

(a) the width of the web plus one fifth of the distance between points of zero moment, or
(b) the actual width of the flange.

(a) will obviously give the smaller value and assuming the distance between points of zero moment to be in the order of 70 per cent of the span AB

$$B = 0.300 + 0.7 \times 11 \times 0.2$$

$$= 1.84 \text{ m}$$

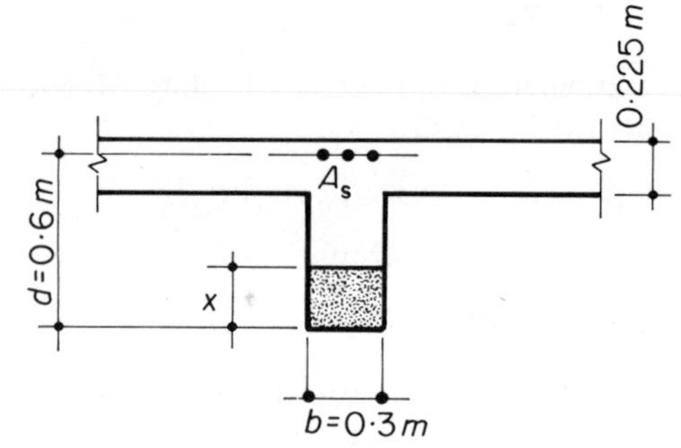

Fig. 5.19

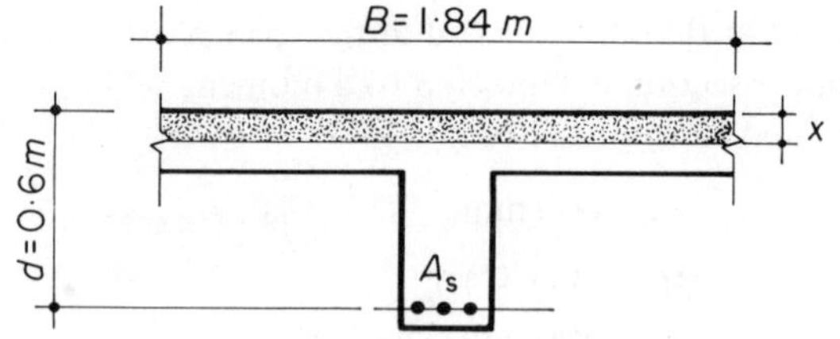

Fig. 5.20

As the neutral axis depth x will be within the slab depth of 225 mm the section may be taken as rectangular with the width b equal to 1840 mm.

$$\frac{M}{bd^2} = \frac{1275 \times 10^6}{1840 \times 600^2}$$

$$= 1.92$$

$$\text{thus } \frac{100A_s}{bd} = 0.58$$

$$A_s = \frac{0.58 \times 1840 \times 600}{100}$$

$$= 6403 \text{ mm}^2$$

Adopt eight 32 mm ϕ (6430).

The maximum shear at A or B will be

$$V = (1.4G_k + 1.6Q_k)5.5$$

ignoring the width of the column

$$\text{thus } V = (1.4 \times 46 + 1.6 \times 21) \times 5.5$$

$$= 98 \times 5.5$$

$$= 539 \text{ kN}$$

$$\text{shear stress } v = \frac{V}{bd}$$

$$= \frac{539 \times 10^3}{300 \times 600} = 2.99 \text{ N/mm}^2$$

$100A_s/bd = 1.42$ thus for $f_{cu} = 30$ N/mm^2 the value of v_c from Table 5 CP 110 is in the order of 0.74 N/mm^2.

Assume 10 mm ϕ links then $A_{sv} = 157$ mm^2 (two legs) and

$$s_v = \frac{0.87 \times 425 \times 157}{300(1.42 - 0.74)}$$

$$= 284 \text{ mm}$$

The spacing of the links should not exceed $0.75 \times d = 450$ mm. Adopt 10 mm ϕ at 275 mm c/c.

To conclude, these examples illustrate that a limit state analysis can involve a considerable increase in arithmetical work with subsequent increase in design office time. With the use of design charts and simplified analytical procedures design time can be reduced, but the danger will always remain that juggling with arithmetic will cloud engineering judgement.

Appendix Reinforcement Areas

APPENDIX

Sectional areas of groups of bars (mm^2)

Bar size (mm)	*Number of bars* 1	2	3	4	5	6	7	8	9	10
6	28.3	56.6	84.9	113	142	170	198	226	255	283
8	50.3	101	151	201	252	302	352	402	453	503
10	78.5	157	236	314	393	417	550	628	707	785
12	113	226	339	452	566	679	792	905	1 020	1 130
16	201	402	603	804	1 010	1210	1 410	1 610	1 810	2 010
20	314	628	943	1 260	1 570	1 890	2 200	2 510	2 830	3 140
25	491	982	1 470	1 960	2 450	2 950	3 440	3 930	4 420	4 910
32	804	1 610	2 410	3 220	4 020	4 830	5 630	6 430	7 240	8 040
40	1 260	2 510	3 770	5 030	6 280	7 540	8 800	10 100	11 300	12 600

Sectional areas per metre width for various bar spacings (mm^2)

Bar size	*Spacing of bars (mm)*								
(mm)	*50*	*75*	*100*	*125*	*150*	*175*	*200*	*250*	*300*
6	566	377	283	226	189	162	142	113	94.3
8	1 010	671	503	402	335	287	252	201	168
10	1 570	1 050	785	628	523	449	393	314	262
12	2 260	1 510	1 130	905	754	646	566	452	377
16	4 020	2 680	2 010	1 610	1 340	1 150	1 010	804	670
20	6 280	4 190	3 140	2 510	2 090	1 800	1 570	1 260	1 050
25	9 820	6 550	4 910	3 930	3 270	2 810	2 450	1 960	1 640
32	16 100	10 700	8 040	6 430	5 360	4 600	4 020	3 220	2 680
40	25 100	16 800	12 600	10 100	8 380	7 180	6 280	5 030	4 190

References

1 Manning G. P., *Reinforced Concrete Design*, Longmans, Green and Co., London, New York, Toronto, first published July 1924.
2 CP 110 : Part 1 : November (1972). *Code of Practice for the Structural Use of Concrete, Part 1, Design, Materials and Workmanship*, British Standards Institution, 2 Park Street, London, W1A 2BS.
3 CP 114 (1957). *The Structural Use of Reinforced Concrete in Buildings*, British Standards Institution, 2 Park Street, London W1A 2BS (reset and reprinted 1965).
4 Cowan H. J., *Reinforced and Prestressed Concrete in Torsion*, Edward Arnold (Publishers) Ltd., 1965.
5 Wood R. H., 'Slab design: past, present and future', Building Research Station Current Paper CP 9/72, Department of the Environment, Garston, Watford, WD2 7JR.
6 Timoshenko S., *Theory of Plates and Shells*, McGraw-Hill, New York, 1959.
7 *ICL 1900 series, Analysis of Plane Frames and Grids*, International Computers Ltd., ICL House, Putney, London, SW15.
8 Wang C. K. and Salmon C. G., *Reinforced Concrete Design*, International Textbook Company, 1965.
9 Bill, Max Robert Maillart. *Bridges and Construction*, Pall Mall Press, 1969.
10 *Travaux, A Half Century of French Prestressing Technology*, Special English Edition, 1966.
11 Comité Europeen du Beton, Bulletin D'Information No. 10, 'Design of structures', January 1959.
12 The Institution of Civil Engineers Research Report, 'Ultimate load design of concrete structures', first published 1962.
13 Comité Europeen du Beton, *International Recommendations for the Design and Construction of Concrete Structures*, June 1970, English Edition, Cement and Concrete Association, London.
14 CP 110 : Part 1 : 1972, *The Structural Use of Concrete*, British Standards Institution, 2 Park Street, London, W1A 2BS.
15 Mitchell G. R., 'Loadings on buildings – a review paper', Building Research Station Current Paper 50/69, Building Research Station, Garston, Watford, WD2 7JR.

16 Jones L. L., *Ultimate Load Analysis of Reinforced and Prestressed Concrete Structures*, Chatto and Windus, 1962.
17 Wood R. H., *Plastic and Elastic Design of Slabs and Plates*, Thames and Hudson, 1961, reprinted 1967.
18 Jones L. L., and Wood R. H., *Yield Line Analysis of Slabs*, Thames and Hudson, Chatto and Windus, 1967.
19 *UNESCO, Reinforced Concrete, An International Manual*, Butterworths, 1971.
20 Comité Europeen du Beton, Bulletin D'Information No. 35 (English translation), 'The application of the yield line theory to the calculations of the flexural strength of slabs and flat slab floors', translated from the French by C. V. Amerongen, issued by the Cement and Concrete Association, London, SW1.
21 *Report of the Inquiry into the Collapse of Flats at Ronan Point, Canning Town*, HMSO, 1968.
22 Ockleston A. J., 'Load tests on a 3-storey reinforced concrete building in Johannesburg', *The Structural Engineer*, Vol. 33, No. 10, October 1955, pp. 304–322.
23 Roberts E. H., 'Load carrying capacity of slab-strips restrained against longitudinal expansion', Building Research Station Current Paper CP 4/70, Building Research Station, Garston, Watford, WD2 7JR.
24 Taylor R., 'A note on a possible basis for a new method of ultimate load design of reinforced concrete slabs', *Magazine of Concrete Research*, Vol. 17, No. 53, December 1965.
25 Wilford M. C., and Yu C. W., 'Catenary action in damaged structures', DOE and CIRIA seminar, The stability of precast concrete structures, London, March 1973.
26 Hillerborg A., 'Theory of equilibrium for reinforced concrete slabs', Betong 41(4), 171 (1956), (translated from the Swedish by C. N. Gibson, Building Research Station Library Communication No. 1082, January 1962).
27 Wood R. H., and Armer, G. S. T., 'The strip method for designing slabs,' Building Research Station Current Paper 39/70, 1970, Building Research Station, Garston, Watford, WD2 7JR.
28 Kemp K. O., 'A strip method of slab design with concentrated loads or supports', *The Structural Engineer*, Vol. 49, No. 12, December 1971.
29 Baker A. L. L., *Ultimate Load Theory Applied to the Design of Reinforced and Prestressed Concrete Frames*, Concrete Publications Ltd., 1956.
30 Baker A. L. L., and Amarakone A. M. N., 'Inelastic hyperstatical frames – analysis and application of the international correlated tests', Presented for discussion at a joint meeting of the Cement and Concrete Association, The Institution of Structural Engineers and The Reinforced Concrete Association in March 1965.
31 Base G. D., and Read J. B., 'The effectiveness of helical binding in the compression zone of concrete beams', Cement and Concrete Association Technical Report TRA/379.
32 Cranston W. B., and Reynolds G. C., 'The influence of shear on the rotation capacity of reinforced concrete beams,' Cement and Concrete Association Technical Report TRA439, 1970.
33 Cranston W. B., 'Analysis and design of reinforced concrete columns', Cement and Concrete Association Research Report No. 20, 1972.
34 *Handbook on the Unified Code for Structural Concrete* (CP 110 : 1972), Cement and Concrete Association, 1972.

35 Taylor R., 'A new method of proportioning stirrups in reinforced concrete beams', Building Research Station Current Paper No. 8, Building Research Station, Garston, Watford, WD2 7JR.
36 Taylor H. J. P., 'Further tests to determine shear stresses in reinforced concrete beams', Cement and Concrete Association Technical Report TRA438, Feb. 1970.
37 Taylor H. J. P., 'Investigations of the forces carried across cracks in reinforced concrete beam by interlock of aggregate', Cement and Concrete Association Technical Report 42.447, November 1970.
38 Regan P. E., and Yu C. W., *Limit State Design of Structural Concrete*, Chatto and Windus, London, 1973.
39 Leonhardt F., and Walther R., *The Stuttgart Shear Tests 1961*, Cement and Concrete Association Library Translation No. 111.
40 Beckett D., *Bridges*, Paul Hamlyn, 1969.
41 Hughes B. P., *Limit State Theory for Reinforced Concrete*, Sir Isaac Pitman and Sons, 1971.
42 Slack J. H., 'Movement joints in concrete structures', Concrete Society Technical Paper PCS 29. Meeting held at University College of South Wales and Monmouth, Cardiff, October 1967.
43 'Control of cracking in concrete structures', *Journal of the American Concrete Institute*, December 1972, ACI committee 224.
44 Base G. D., Read J. B., Beeby A. W., and Taylor H. P. J., 'An investigation of the crack control characteristics of various types of bar in reinforced concrete beams', Cement and Concrete Association Research Report 18, Part 1.
45 Beeby A. W., 'An investigation of cracking in slabs spanning one way', Cement and Concrete Association Technical Report TRA433, 1970.
46 Farebrother J. E. C., 'Designers' point of view of CP 110', The Institution of Structural Engineers/Concrete Society, Unified Code Symposium, Cafe Royal London, September 1973.

Index

Aggregate interlock, 73
American Concrete Institute, 31
Amplification of moment, 18, 65
Anchorage bond, 12, 79

Baker, A. L. L., 61
Beams,
  bending moment coefficients (elastic), 19
  elastic analysis for: bond, 11
    flexure, 2
    shear, 8
    torsion, 12
  flexural rigidity, 6
  moment redistribution in, 59
  plastic analysis, 61
  T & L, 71
Bond,
  anchorage, 12, 79
  local, 12, 78

Characteristic,
  loads, 33
  strengths, 32
Columns,
  amplification of moment, 18, 65
  axial load, 79
  axial load plus bending, 80
  calculations, 119
Concrete,
  characteristic strength, 32
  design strength, 33
  stress–strain relationship, 24, 69
Crack widths,
  calculations for, 102
  formulae, 85
  permissible values, 89
Creep, 7, 94, 108

Deflections,
  calculations for, 106
  formulae, 91
  limiting span to depth ratios, 89, 90
Doubly reinforced beams, 6, 70

Elastic theory,
  beams, 15
  bond, 11
  flexure, 2
  shear, 8
  slabs, 17
  torsion, 18

Flat slabs,
  analysis, 52
  calculations, 112
  punching shear, 76

Flexural rigidity, 6
Frame analysis, 18
Freyssinet, E., 24

Hillerborgs method, 54
Hinge rotation calculations, 63–65
Hinges, plastic, 61

Limit state,
  development of theory, 23
  serviceability, 36, 84
  ultimate, 34, 40
Local bond, 12, 78

Maillart, R., 84
Manning, G. P., 1
Membrane effects, 54
Moment/curvature, 27
Moment redistribution, 56

Neutral axis,
  elastic theory, 3
  restrictions on depth, 59
  tabulated values, 4

Parabolic rectangular stress block, 68
Partial safety factors, 33
Plastic theory,
  beams, 61
  slabs, 40
  torsion, 76
Punching shear, 76

Rectangular stress block, 70
Redistribution of moment, 59

Reinforcement,
  sectional areas, 126, 127
  stress–strain relationship, 71

Serviceability,
  cracking, 85
  deflection, 89
  general, 84
Shear,
  in beams, 73
  in slabs, 75
  reinforcement for, 74
Slabs,
  elastic analysis, 17
  flat, 52
  Hillerborgs strip method, 54
  yield line analysis, 40
Stress–strain relations
  for concrete, 24, 69
  for reinforcement, 71
Strip method, 54

Torsion,
  in beams – elastic, 12
    plastic, 76
  reinforcement for, 78

Under-reinforced beams, 26

Wood, R. H., 40, 51

Yield line theory,
  corner effects, 45
  development of, 40
  equilibrium method, 43
  work method, 44